NEW TOPICS IN MONOMER AND POLYMER RESEARCH

NEW TOPICS IN MONOMER AND POLYMER RESEARCH

ALBERTO D'AMORE
GENNADY ZAIKOV
EDITORS

Nova Science Publishers, Inc.
New York

For permission to use material from this book please contact us:
Telephone 631-231-7269; Fax 631-231-8175
Web Site: http://www.novapublishers.com

Library of Congress Cataloging-in-Publication Data

New topics in monomer and polymers research / Alberto D'Amore and Gennady Zaikov.
p. cm.
Includes index.
ISBN 13 978-1-60021-436-3
ISBN 10 1-60021-436-3
1. Monomers--Research. 2. Polymers--Research. 3. Polymerization--Research. 4. Chemical reactions--Research. I. D'Amore, Alberto. II. Zaikov, Gennadii Efremovich.
QD281.P6N49 2006
547'.7--dc22 2006028915

Published by Nova Science Publishers, Inc. ✦New York

CONTENTS

PREFACE

This book is mainly dedicated to the processes of ozonolysis, destruction and stabilization of polymers, reactions and structure formation of polymers in fractal spaces, hydrosilylation reactions, antiadhesive coatings, synergism and antagonism in processes of polymers phototransformation and light stabilization, selective oxidation, carbocations reactions and polymer analogous reactions.

The design of barbotage-rotation installation with different profiles of blades of the rotator has been created on the basis of modeling method. As a result of the complex of aerohydrodynamic research in the system "gas - liquid" carried out on the above unit certain recommendations on the improvement of the construction and operating conditions have been developed with a view to a rise in the effectiveness of dust removal. Barbotage-rotation installations designed for clearing and cooling gas discharges are applied in chemical, oil refining, and other fieds of industry. As discussed in Chapter 1, the above units are important in solving ecological problems, e.g. the refining process of waste (exhaust) gas removed from the unit into the atmosphere. That's why research on the improvement of the design and operating conditions of barbotage-rotation units has always been urgent and remains a problem today. With a view to intensification of the refining process and a rise in the effectiveness of the barbotage-rotation unit, a complex of aerohydrodynamic research on the improvement of the construction of the rotator and operating conditions of the unit has been carried out.

The reaction of hydride polyaddition of methylhydridesiloxane an methylhydridesiloxane-dimethylsiloxane oligomers to unsaturated carbocyclic compound - tricyclodecadiene, at the various ratio of initial compounds, in the presence of platinum hydrochloric acid has been investigated. It was shown that hydrosilylation reaction proceeds both to "1-2" and "9-10" position by intermolecular interactions with formation of branching systems. The polyorganosiloxanes with tricyclodecenyl fragment in the side chain, completely soluble in organic solvents were synthesized. For full characterization of hydride addition of methylhydridesiloxane to tricyclodecadiene, by quantum-chemical half empiric AM1 method, for all initial, intermediate and final products, in modelling reaction of hydrosilylation of methyldimethoxysilane to tricyclodecadiene, the heats of formations (ΔH_f), change of energy (ΔH) of the system depending on the change of distance ($R_{C\text{-}Si}$) between ≡C-Si≡ bonds, also the charges values (q) on the atoms, dipole moments (μ) and bonds orders (Pij) are calculated. It was concluded in Chapter 2, that the course of hydride addition of modelling reaction of methyldimethoxysilane to tricyclodecadiene energetically is more favourable by 9.10-addition. The hydride polyaddition reaction order, activation energies and rate constants were found. The synthesized oligomers were characterized by ^{1}H and ^{13}C NMR, FTIR, thermogravimetry, gel permeation chromatography, differential scanning calorimetry and wide angle X-ray methods.

Two series of polymers having different tacticity-dependent stereochemical micro-structures have been submitted to controlled ageing. One series consists of three PVC samples prepared at distinct temperatures and of the same samples after substitution reaction to various degrees. And one series consists of three commercial samples of PMMA. In both cases a straight correlation between enthalpy loss during ageing and the content of mmr tetrads termini of isotactic sequences of at least one heptad in length is found, especially with those mmr structures that take the GTTG^{-}TT conformation. In the case of PVC that is more quantitavely evidenced through the steroselective removal of this conformation as the result of a substitution reaction. Anyway, the results shown in Chapter 3 shed light on the mechanisms at the nanometric level of the ageing process in polymers.

Approaches to the consideration of a coal swelling process, which were used up to now and based on the theory of regular solutions, do not give the possibility to generalize quantitatively the experimental data. As discussed in Chapter 4, adequate relation between the physical–chemical properties of the solvents and the degree of a coal swelling in them can be obtained only with the use of linear multiparametric equations which take into account the effects of the all processes proceeding in the system; besides, the basicity and a molar volume of the liquids are determinative. Such approach is effective at the generalization of data concerning to extraction of a coal.

In Chapter 5, reciprocal influence of antioxidizing and detergent-dispersive components of batch additives have been studied. Oligomers of piperylene and isobutylene functionalized by element sulfur and phenols in combination with succinimide have been chosen as the object of research. Constants of inhibition rates have been determined for oligomer antioxidizing agents.

There are several definitions used for description of biological destruction. As discussed in Chapter 6, they have no significant difference, all saying that biodegradation is a process caused by biological activity that leads to generation of natural metabolic end-products with the change in chemical structure of material.

In Chapter 7, the present state of the synthesis of cyclic organosiloxanes and organocarbosiloxanes with assigned disposition of various functional groups is reviewed. Organosilicon monomers containing functional groups at silicon atoms are extremely inclined to various reactions of transformation, which produce ≡Si-O-Si≡ bonds. At the present time, the most practically significant organosilicon polymers are industrially produced from organochlorosilanes of various functionalities. To carry out directed synthesis of oligomers and block-copolymers with cyclic and polycyclic fragments in the macromolecular backbone, methods of linear di- and poly-functional oligochlororganosiloxanes synthesis have been developed, on the basis of which or-ganocyclosilanes with the given disposition of chlorine atoms and other functional groups at atoms of silicon may be derived.

There is some information in Chapter 8, about synthesis of functional group containing organocyc-losiloxane. Synthesis of organocyclosiloxane with different disposition of functional groups at atoms of silicon is mainly reduced to two methods: hydrolytic alkoxyorganosilanes and hetero functional condensation (HFC) reaction of silanes with siloxanes containing functional chlorine-, alkoxy-, amino and hydroxyl groups or to modification of previously prepared organocyclosiloxanes. The first series of works has used co-hydro-lysis products of dimethyldichlorosilane with phenyltrichlorosilane, dimethyldichlorosilane with methyltrichlorosilane or methylphenyldichlorosilane with phenyltrichlorosilane. During these reactions, organocyclosiloxanes with low yields and statistical arrangement of functional groups at silicon are obtained. From these two methods of synthesis of the functional group containing organocyc-losiloxanes, the most acceptable reaction is HFC.

The reaction of hydrosilylation of α,ω–bis(trimethylsiloxy)methylhydridesiloxane ($n\approx53$) with acrylic acid at various ratios of initial compounds, in the presence of Platinum hydrochloric acid has been investigated. It was established in Chapter 9, that in the presence of Platinum hydrochloric acid competitive dehydrocondensation reactions take place with formation of oligomers with various link of chain. In high stages of reaction, three-dimensional systems were obtained. During dehydrocondensation the reaction order, reaction rate constants and activation energy were found. The structures of synthesized oligomers were determined by IR and NMR spectra data. Gel permeation chromatographic, differential scanning calorimetric and X-ray analysis of synthesized oligo-mers were carried out.

The influence of 5,10,15,20-tetrakis-(3,5-di-tert-butylphenyl)porphyrin and its metal complexes on the radical polymerization of methylmetacrylate initiated by benzoyl peroxide has been studied. It was revealed that metal porphyrins form an initiating system with a peroxide initiator and increase the rate of the initial stage of polymerization. It was shown in Chapter 10, that addition of porphyrins leads to the formation of polymethylmetacrylate with lower molecular weight, enhanced thermostability and higher stereoregularity due to the increased presence of syndiotactic sequencies.

The free-radical polymerization of methyl methacrylate has been investigated in the presence of N, N-dimethyl-N - (methylferrocenyl)amine (FCAM). The mutual influence of ferrocenyl- and amine - groups on kinetic parameters of polymerization and on the microstructure of received polymers has been found. It is shown in Chapter 11, that poly-methyl methacrylate obtained in the presence of FCAM has mainly sindiotactic structure.

In Chapter 12, works on the synthesis and research into fullerene-containing polymers (FP) are paid particular significance in the chemistry of fullerenes. Radical polymerization is seen as especially attractive with such aims in view due to relative simplicity of the process to be realized, availability and variety of starting substances. The ways of obtaining fullerenes by vinyl monomer-based radical polymerization have been mainly discussed. However up to date there are no reports on synthesis of these high-molecular compounds from the allyl order monomers. Interest is taken in the studies of the effect of fullerenes on the polymerization of the monomers containing simultaneously both vinyl and allyl double bonds, as well as the compounds having only allyl group. They have been first to do research into radical (co)polymerization of allyl ester of metacrylic acid (AMA) and allyl chloride (AC) in the presence of C_{60} fullerene.

In: New Topics in Monomer and Polymer Research
Editors: A. D'Amore and G. Zaikov, pp. 1-7 ISBN: 1-60021-436-3

Chapter 1

OF COMPLEX AERODYNAMIC RESEARCH AND THE EFFECTIVENESS OF ARRESTING DISPERSED PARTICLES FOR BARBOTAGE – ROTATION

A.K. Panov, R.R. Usmanova, V.G. Zaikov* and G.E. Zaikov*

Branch UGNTU, 2 Oktyabrya bevd., Sterlitamak 453118, Bashkortostan, Russia
*N.M. Emanuel Institute of Biochemical Physics, RAN, 4 Kosygin Street, Moscow 119991, Russia

ABSTRACT

The design of barbotage-rotation installation with different profiles of blades of the rotator has been created on the basis of modeling method.

As a result of the complex of aerohydrodynamic research in the system "gas - liquid" carried out on the above unit certain recommendations on the improvement of the construction and operating conditions have been developed with a view to a rise in the effectiveness of dust removal.

Barbotage-rotation installations designed for clearing and cooling gas discharges are applied in chemical, oil refining, and other fieds of industry.

The above units are important in solving ecological problems, e.g. the refining process of waist (exhaust) gas removed from the unit into the atmosphere. [1]

That's why research on the improvement of the design and operating conditions of barbotage-rotation units has always been urgent and remains problem of today. [2]

With a view to intensification of the refining process and a rise in the effectiveness of the barbotage-rotation unit, a complex of aerohydrodynamic research on the improvement of the construction of the rotator and operating conditions of the unit has been carried out.

1. Description of the Unit for Dust Removal Process Research and the Technique (Method) of the Experiment

Research on aerohydrodynamics of curled (rotated) flows and the effectiveness of dust removal in the system "gas - liquid" was carried out on the barbotage-rotation unit with different profiles of blades of the rotators.

The unit was designed taking into consideration possible approximations to real conditions of industrial installations. (Fig. 1)

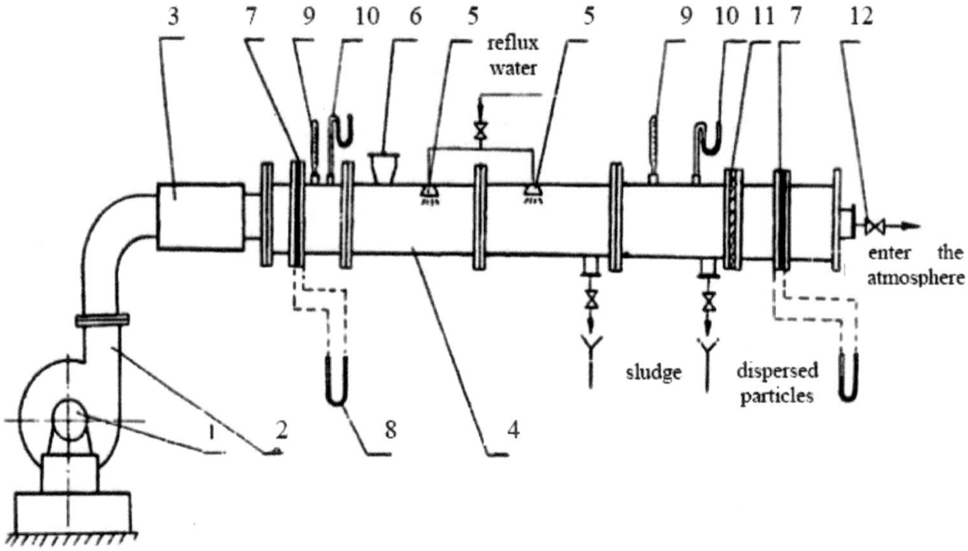

Figure 1. Flow diagram of the experimental unit; 1 – electric motor; 2 – ventilator (fan); 3 – electric heater; 4 – rotation chamber (cell); 5 – sprayers; 6 – bunker; 7 – diaphragm; 8 – pressure – gauge; 9 – thermometer; 10 – manometer; 11 – filter; 12 – valve.

It was taken into account that the rotator should ensure not only a high degree of dust removal but be constructively well coordinated with the barbotage – rotation device.

Air and white soot (smoke - white) with the particle size of d = 2 ÷ 40 mkm were used as a model system.

Atmospheric air forced by ventilator 2 (VC 1446, producer being Industrial Ventilation, Sterlitamak, Bashcortostan) was passed throngh electric heater 3 at the required heat temperature (100^0C). Dusty air rate of flow $G=(0,6\div4,0)\cdot10^{-2}$ m^3/s. After reaching the operating temperature conditions and sprayers 5, placed in front of and behind the rotator, the flow was given reflux liquid, reflux feed being regulated by rotameter RS – 7. Simultaneously the flow was given white soot from feed bunker 6. While the dusty air was passing through the blade rotator, in the field of centrifugal forces dust removal and cooling of the gas flow took place. The unremoved dispersed particles of the flow fell out in filter 11. The removal of the sludge and dispersed particles was carried out through unions in the bottom of the device.

The purified air was discharged into the atmosphere.

Sprayers 5 were placed in front of and behind the atmosphere rotator for the improvement of purifying conditions.

Reflux feeding was regulated by rotameter RS – 7 ($Q=(5,5\div14)\cdot10^{-4}m^3/s$).

The liquid reflux favoured the enlargement of the unremoved, dispersed particles and the reduction of time of removal. The sludge formed was washed off with the liquid sprayed by the sprayer placed behind the rotator and dischared into the sludge – collector.

The subject of the Research was constructions of rotators with four kinds of profiles of blades: parabolic, hyperbolic, sine wave and direct (Fig. 2).

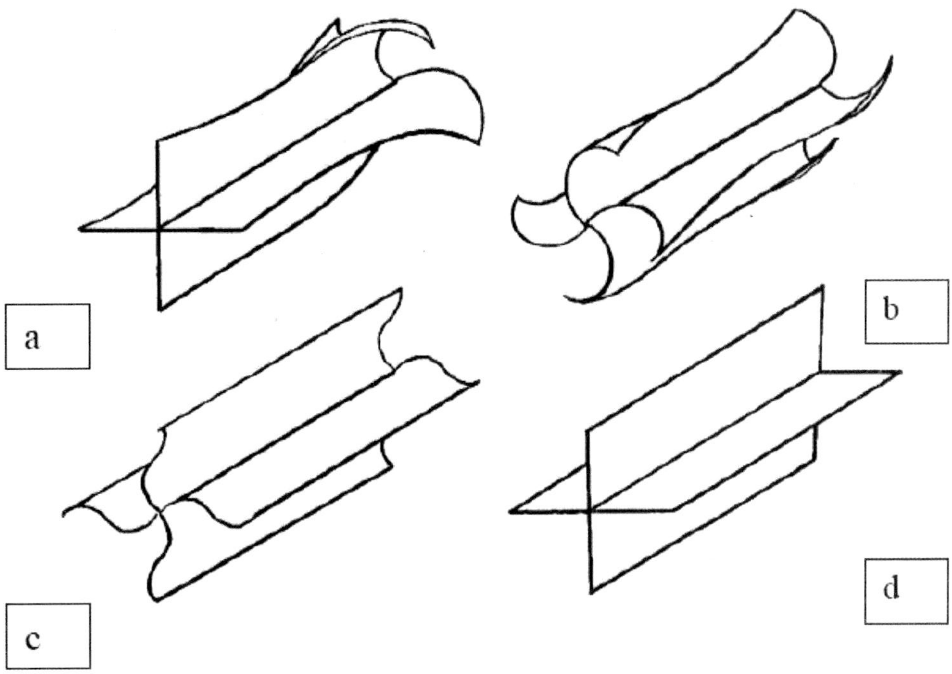

Figure 2. Profiles of blades of rotators; a – parabolic; b – hyperbolic; c- sine wave; d – direct.

2. Evaluation of the Effectiveness of Dust Removal

Full (complete) effectiveness of dust removal was (calculated) expressed in percentage [3] as a ratio between mass of removed dust and mass of initial dust.

$$\eta = \frac{C_n - C_k}{C_n} = \frac{C_{rem}}{C_n} \cdot 100\%, \qquad (1)$$

where C_n – mass expenditure of rigid particles on exit, kg / s;
C_{rem} – mass expenditure of removed particles, kg / s.

According to the results of the effectiveness accouts there were built diagrams for different profiles of blades of the rotator. As the data obtained shows for the models of blades of the rotator being under research the main mass of particles is removed completely. The effectiveness of dust removal remains within 85 – 99 %. With high reflux common efficiency of the device rises (Fig. 3). With minimum reflux of Q = 8,5 · 10^{-4} m^3/s, common efficiency comes to 85 %.

The data obtained shows that even with little reflux a high degree of dust removal is obtained. Research shows that in the investigated range the initial dust concentration practically does not influence the effectiveness of dust removal. This is of great importance for the device work in industrial service conditions where the dust content being the function of the technological process conditions may vary over a wide range.

With high air expenditure the device efficiency falls (drops) (Fig. 4). This can be explained by the fact that at significant air expenditure and little water expenditure some rigid particles have no time to enter the contact with the liquid and are discharged from the device.

In the diagram points (dots) show experimental efficiency values, continuous lines show theoretical values calculated by formula (1).

It is interesting to note that the least dust removal efficiency is obtained for rectangular profile of blades, coming to 85 %.

The highest device efficiency (up to 99 %) is obtained for parabolic profile of rotator blades.

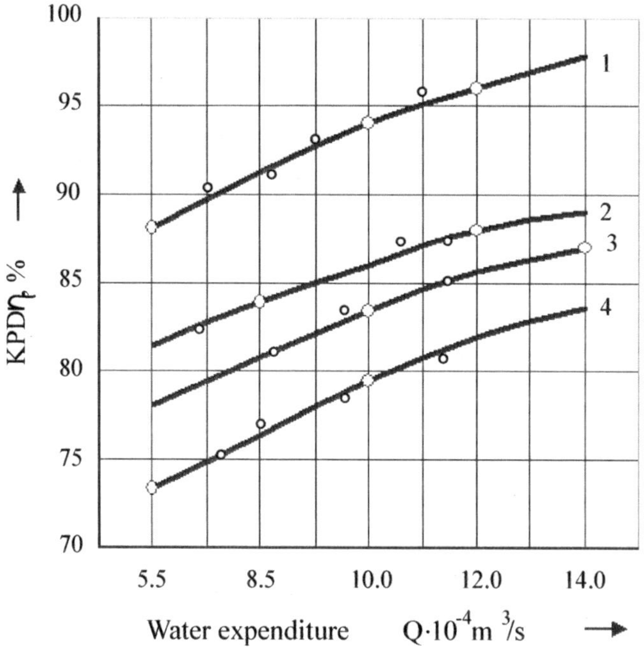

Figure 3. Dependence of the device efficiency on water expenditure at constant air expenditure (G = 40 · 10^{-3} m^3 / s) for different profiles of the rotator blades; 1 – parabolic; 2 – hyperbolic; 3 – sine wave; 4 – direct.

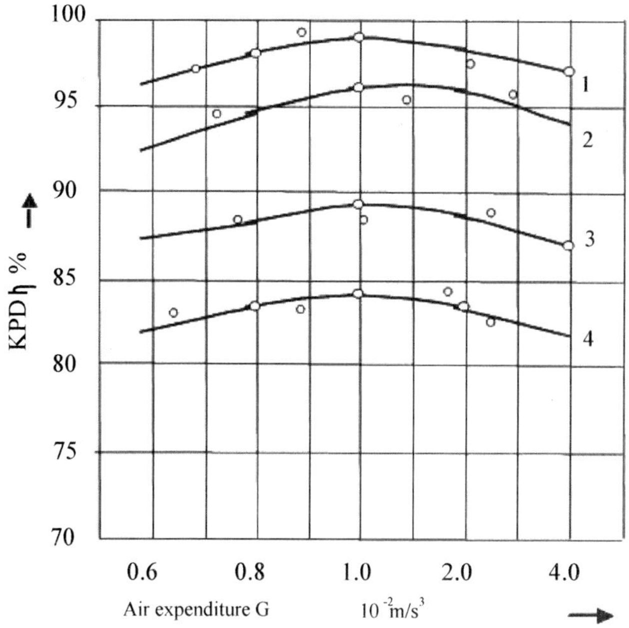

Figure 4. Dependence of the device efficiency on air expenditure at constant water expenditure ($Q=8{,}5 \cdot 10^{-3} m^3/s$) for different profiles of the rotatot blades; 1-parabolic; 2- hyperbolic; 3- sine wave; 4 – direct.

3. Aerohydrodynamic Research Results for Barbotage – Rotation Device in Comparison with Theoretical Ones

There is a general tendency to a rise in hydrolic resistance with high speed in a free cut.

The maximum value of hydrolic resistance at specific reflux of $Q = 14 \cdot 10^{-4}\ m^3 / s$ comes to $\Delta P = 30$ Па for hydrolic profile of blades, $\Delta P = 20$ Па for direct and sine wave ones.

Thus constructive modifications of the profiles of the rotator blades enable one to raise significantly dust removal effectiveness (Fig. 3, 4.) with slight (small) hydrolic resistance increase.

Curling efficiency of rotator K varies from 0,2 ÷ 1,0. The experimental data analysis shows that hydrolic resistance efficiency ξ at some value K does not depend on liquid expenditure which is explained by two factors connected with reflux feeding of the device.

On the one hand reduction of ξ is observed because of gas tangential speed drop on the account of brake action of the liquid, on the other hand increase of ξ is connected with a rise in gas flow pressure on the liquid carrier.

On this foundation ξ account technique was developed including ξ account formula for the "dry" device

$$\xi_{dry} = \frac{1}{n}((R/r)^{2n} - 1) + \frac{1}{K^2}\left(\frac{V_2}{V_1}\right)^2, \tag{2}$$

the formula of loss account of gas pressure on the liquid carrier

$$\xi_{tr} = 4 \cdot \left(\frac{Q}{G}\right)^{0,6} \sqrt{1 + \frac{1}{K^2}}, \tag{3}$$

and the final formula for ξ account of the reflux fed rotation device

$$\xi_{sp} = \frac{1}{n}((R/r\ )^{2n} - 1) + \frac{E^2}{K^2}\left(1 + \frac{\rho_l}{\rho_g}\right) \cdot \left(\frac{V_2}{V_1}\right)^2 + 4\left(\frac{Q}{G}\right)^{0,6} \sqrt{1 + \frac{1}{K^2}}, \tag{4}$$

taking into consideration the presence of a disperse phase and partial loss of the flow curling.

The ξ accounts by this formula will be well coordinated with the experiment data (Fig. 5).

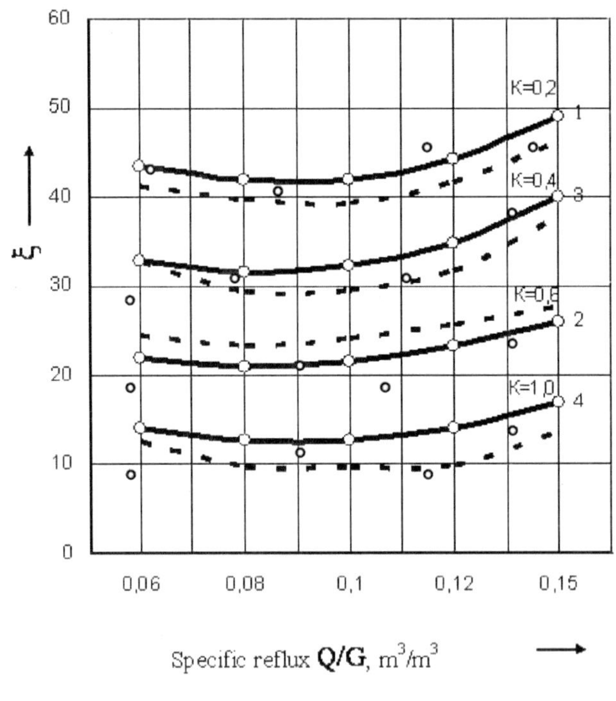

Figure 5. ξ dependence on specific reflux of the device with different values of curling efficiency.

In the diagram theoretical values of ξ are marked by a continuous line, the experiment data is marked by points (dots). The relative divergence of theoretical values calculated by formula (4) and the experimental data made no more than 15 %.

Thus the technique offered with an adequate accuracy of accounts can be recommended for aerohydrodynamic research of the flows in the barbotage – rotation device with different constructive parameters of the rotator.

Conclusion

1. Taking into consideration research results on dust removal from gas it is advisable (expedient) in what follows to use the barbotage – rotation device with parabolic profiles of rotator blades, having high effectiveness of dust removal (up to 99 %) and small hydrolic resistence, as an optimum variant in dust removal system.
2. The introduction of barbotage – rotation device with parabolic profile of rotator blades is offered to Bashcortostan refinery plants for dust removal and cooling gas discharges.

Conditional Designations

E	loss efficiency of curling
G	gas expenditure, m^3 / s
K	curling efficiency, k = (0,2 ÷ 1,0);
n	experiment constant;
Q	liquid expenditure, m^3 / s;
R	apparatos radius, m;
r	rotation radius ,m
V_1	gas speed on inlet, m / s;
V_2	gas speed on inlet, m / s;
P_g	gas density, kg / m^3;
P_l	liquid density, kg / m^3;

References

[1] Patent 2182843 RF MKI V 01 D 47 / 06 Barbotage – *rotation device* / A.K. Panov, R.R. Usmanova, V.M. Titov et al. – N. 200012 / 12; requested 27.11.00; published 27.05.02.

[2] Panov A.K., Usmanova R.R., Minsker K.S. Effectiveness raise of the process of clearing furnace combustion gases / *Chemical Industry* / - 2003 N. 4. P. 19 – 24.

[3] Ugov V.N., Valdberg A.J. Preparation of industrial gases for clearing. – M.: *Chemistry*, 1985. – P. 206.

In: New Topics in Monomer and Polymer Research
Editors: A. D'Amore and G. Zaikov, pp. 9-22
ISBN: 1-60021-436-3

Chapter 2

SYNTHESIS AND INVESTIGATION OF PROPERTIES OF NEW THERMOREACTIVE POLYSILOXANES

O. Mukbaniani**[1], G. Zaikov, T. Tatrishvili, N. Mukbaniani and Kh. Koberidze

[1] Ivane Javakhishvili Tbilisi State University,
Ilia Chavchavadze Avenue, 1, Tbilisi 0128, Georgia
2 Institute of Biochemical Physics, Russian
Academy of Sciences, 4 Kosygin Street, Moscow 119991 Russia

ABSTRACT

The reaction of hydride polyaddition of methylhydridesiloxane and methylhydridesiloxane-dimethylsiloxane oligomers to unsaturated carbocyclic compound - tricyclodecadiene, at the various ratio of initial compounds, in the presence of platinum hydrochloric acid has been investigated. It was shown that hydrosilylation reaction proceeds both to "1-2" and "9-10" position and by intermolecular with formation of branching systems. The polyorganosiloxanes with tricyclodecenyl fragment in the side chain, completely soluble in organic solvents were synthesized. For fully characterization of hydride addition of methylhydridesiloxane to tricyclodecadiene, by quantum-chemical half empiric AM1 method, for all initial, intermediate and final products, in modelling reaction of hydrosilylation of methyldimethoxysilane to tricyclodecadiene, the heats of formations (ΔH_f), change of energy (ΔH) of the system depending on the change of distance (R_{C-Si}) between ≡C-Si≡ bonds, also the charges values (q) on the atoms, dipole moments (μ) and bonds orders (Pij) are calculated. It was concluded, that the course of hydride addition of modelling reaction of methyldimethoxysilane to tricyclodecadiene energetically is more favourable by 9.10-addition. The hydride polyaddition reaction order, activation energies and rate constants were found. The synthesized oligomers were characterized by ^{1}H and ^{13}C NMR, FTIR, thermogravimetry, gel permeation chromatography, differential scanning calorimetry and wide angle X-ray methods.

Keywords: organosiloxanes, hydride polyaddition, oligomers, thermal-oxidative stability.

---

* Corresponding author: mukbaniani@ictsu.tsu.edu.ge

INTRODUCTION

The properties of siliconorganic polymers depend as on the structure of a macromolecular chain and of the nature of the surrounding organic groups at silicon atom [1].

Therefore it's evident that effective methods of synthesis of new type siliconorganic co-polymers with complex's valuable properties attach a great importance. Particularly the synthesis of polyorg-anosiloxanes with unsaturating surrounding groups at silicon atom, for receiving of siliconorganic rubbers [2] and for modification carbochain polymers [3] is very interesting.

Macromolecular grafting is a synthetic approach that involves the functionalization of a preformed polymer backbone containing reactive groups with an appropriate reagent. There is an increasing interest in catalytic grafting of polyhydrosiloxane as to an alternative route for the synthesis of organic–inorganic hybrid polymers, owing to the complexities involved in direct generation of well-defined linear hybrid polymers [4]. Besides the process of equilibration, the hydrosilylation process is the most widely used method of preparing organofunctional polymethylsiloxanes starting from poly(methylhydro)siloxanes [5, 6]. From this point of view the chemical modification reactions are very interesting. In literature the modification reactions of the linear polymethylvinylsiloxanes with cyclopentadiene groups is known [7]. In this case the synthesized polymers are characterized with higher reaction ability for vulcanization and thermal-oxidative stability.

From literature [8] it's known, that electrophilic addition of monomer type methylhydrochlorosilanes to tricyclodecadiene in the presence of Spier catalyst proceeds both to 9.10 and 1.2 addition depending on the nature and amount of organic groups at silicone atoms, according to the following scheme:

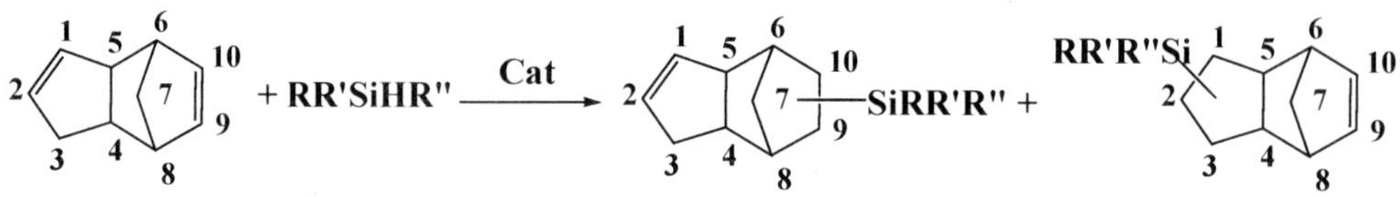

Where: R=R'=R"=Alk, Ar; R≠R'≠R"≠Alk, Ar; R"=Cl.

By the authors [9] it was shown that using in this reactions such catalysts as $Rh_4(CO)_{12}$ or $Co_2(CO)_8$ the hydrosilylation proceeds selective mainly via 9,10-addition, without obtaining isomeric adducts.

2. EXPERIMENTAL PART

2.1. Materials and Techniques

The starting materials for the synthesis of thermoreactive comb-type methylorganosiloxane oligomers with functional groups in the side chain were α,ω–bis(trimethylsiloxy)methylhydrosiloxane and tricyclodecadiene.

The initial α,ω–bis(trimethylsiloxy)methylhydridesiloxane with degree of polymerization ~m≈30, 53 and α,ω–bis(trimethylsiloxy)methylhydridesiloxane-dimethylsiloxane oligomer with degree of polymerization ~m≈33, n≈23 tricyclodecadiene and catalyst H_2PtCl_6 were obtained from Sigma-Aldrich and used as received. The organic solvents were cleaned by drying and distillation.

The organic solvents were cleaned by drying and distillation. The purity of initial dicyclopentadiene was controlled by its boiling temperature and refraction indexes. The purity of the starting compounds was controlled by gas-liquid chromatography "LKhM-8 MD", (Russian model) phase SKTF-100 (10%), the NAW chromosorb, carrier gas He, the 2M column.

FTIR spectra were obtained on Nicolet Nexus 470 with MCTB detector. ^{1}H and ^{13}C spectra were recorded on AC-250 spectrometer at 250 MHz in the solution of deuterium solvents as solvent and internal standard.

Thermogravimetric analysis was performed on a Perkin-Elmer TGA-7 the thermogravimetric analyser at ~10 ^{0}C/min. Differential scanning calorimetric analysis was performed on a Perkin-Elmer DSC-7. Heating and cooling scanning rates were ~5 ^{0}C/min.

Gel-permeation chromatography (GPC) investigations were performed using of Waters Model 6000A chromatograph with an R 401 differential refractometer detector. The column set comprised 10^3 and 10^4 Å Ultrastyragel columns (Waters, USA). Sample concentration was approximately 3% by weight in toluene; a typical injection volume for the siloxane was 5 μL. Standardization of the GPC was accomplished by the use of styrene or polydimethylsiloxane standards with the known molecular masses.

Wide-angle X-ray diffractograms were taken on a DRON-2 (Burevestnik, Saint-Petersburg, Russia) instrument using A-CuK_α radiation. The values were measured without a filter, and the angular velocity of the motor was ~2 deg/min.

2.2. Hydride Addition of Methylhydridesiloxane to Tricyclodecadiene (TCD)

The hydrosilylation was carried out in the three-necked flask equipped with stirrer and reflux condenser with calcium chloride tube. The initial reagents, α,ω–bis(trimethylsilyl)methylhydridesiloxane and tricyclodecadiene were placed in the flask and thermostated in an oil bath until constant temperature was achieved. As a catalyst 0.1 M solution of platinum hydrochloric acid in tetrahydrofuran ($5\div10\times10^{-5}$ g per 1g of the starting substance) was used. The reaction was performed in the argon atmosphere in the presence of toluene at temperature region 60÷90°C. Then the reaction mixture was connected to the vacuum (P=1÷2 mmHg) and the toluene and unreacted TCD was removed in the temperature range 40-50^{0}C under the vacuum.

3. Results and Discussion

In literature there is some information of receiving tricyclodecenil containing alkyl siloxanes by hydrosilylation of TCD by hydrideorganochlorosilanes in the presence of platinum hydrochloric acid [7], in the presence cobalt and rhodium carbonyls [8]. It's known the polycondensation type organo-inorganic copolymers with tricyclodecenil groups in the chain [9]. These above mentioned copolymers with unsaturated fragments in the chain by practical stand-point are very interesting systems, because they contain unsaturated bonds, which give us possibility to change the properties of obtained siliconorganic elastomers in a wide range.

Preliminary heating of initial compounds in the temperature range of 60÷90^{0}C in the presence of catalyst, showed that in these conditions polymerization of tricyclodecadiene, break in siloxane backbone, or elimination of methane do not take place. Besides, there are no changes in the NMR and FTIR spectra of TCD and methylhydridesiloxane. By gas-liquid chromatography it was established that the polymerization of TCD in this condition does not proceed.

For the purpose of synthesis of the linear siliconorganic elastomers with tricyclodecenil fragments in the side chain the modification reactions of α,ω-bis(trimethylsiloxy)methylhydridesiloxane (n≈ 30,53) and α,ω-bis(trimethylsiloxy)methylhydridesiloxane-dimethylsiloxane with TCD were studied. The hydrosilylation reactions were studied at 1:30, 1:53 and 1:25 molecular ratio of initial compounds in the presence of platinum hydrochloric acid as a catalyst. As it's seen from the molecular structure of TCD:

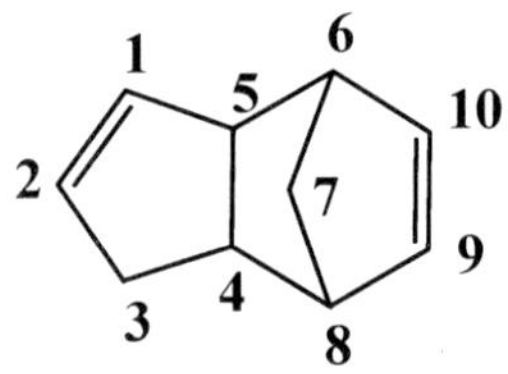

it contains two unsaturated bonds in "1-2" and "9-10" position. From literature data [8,9] during hydrosilylation of TCD in case of a big amount of electroacceptoric groups containing organochlorosilanes (methyldichlorosilane) hydride addition at double bond proceeds to "9-10" position. At a small amount of electroacceptoric groups (in case of trimethylchlorosilane) the addition proceeds both to "1-2" and "9-10" position. In the presence of electro donor groups (in case of trimethylsilane) the hydrosilylation mainly proceeds to "1-2" position. It was possible that in our case the reaction must be proceeding to "1-2" position. As it's known from literature the electrophilic addition of ≡Si-H bond to TCD must be proceed with the formation of transitional complex. The stability of the complex at "9-10" position because of steric difficulty is lower, than the stability of the complex to "1-2" position, because of that the addition mainly must be proceeded to "9-10" position.

Table 1. ^{1}H NMR resonance signals of TCD and oligomers (II, III)

C-H bond	H-1	H-2	H-3	H-4	H-5	H-6	H-7	H-8	H-9	H-10
TCD (ppm)	5,95	5,95	2,20	1,67	1,67	2,22	1,40	2,17	5,50	5,50
II (ppm)	5,95	5,95	2,20	1,70	1,70	1,80	1,30	1,80	-	-
III (ppm)	5.96	5,96	2,20	1,70	1,70	1,80	1,30	1,80	-	-

In the ^{1}H NMR spectra of TCD, Figure 1 one can observe the multiplet resonance signals for unsaturated bonds with center δ=5.50 ppm and δ=5.95 ppm characteristic of unsaturated bonds. One can observe for protons in methylene groups at 7a and 7b position with center δ=1.55 ppm and δ=1.35 ppm (J_{hemin}=14 Hz) $\delta H(3_{a,e})$=2.20 ppm.

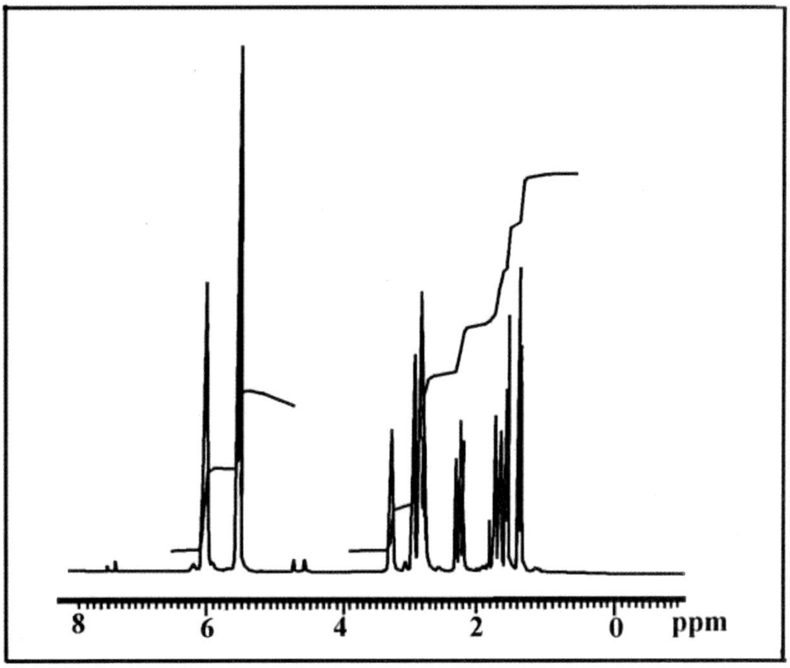

Figure 1. ^{1}H NMR spectrum of tricyclodecadiene.

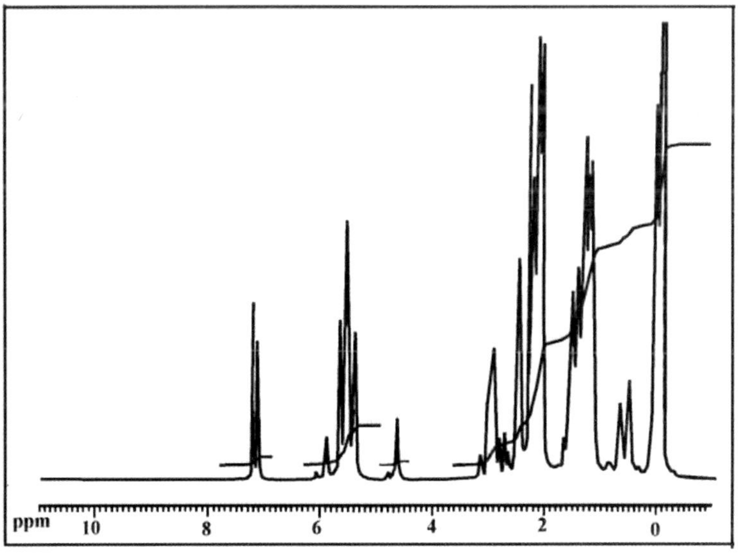

Figure 2. ^{1}H NMR spectrum of α,ω-bis(trimethylsiloxy)methyltricyclodecenilsiloxane II.

In the ^{1}H NMR spectra of hydrosilylation product II one can observe the resonance unimportant signals for protons in unsaturated fragment at "9-10" position, and mainly unchanged multiplet resonance signals at δ=5,3÷5,7 ppm, with integral intensity ~1:10. So, from ^{1}H NMR spectra of hydrosilylation product it's evident that the hydride addition mainly proceeds at "9-10" position and about 10% in "1-2" position.

The above-mentioned conclusion is in conformity with quantum-chemical calculation. By Quantum-mechanical semi empiric method AM1 (Austin Model) the electronic characteristics: the charges on the atoms and bonds orders were calculated. The obtained results are presented on the diagram 1.

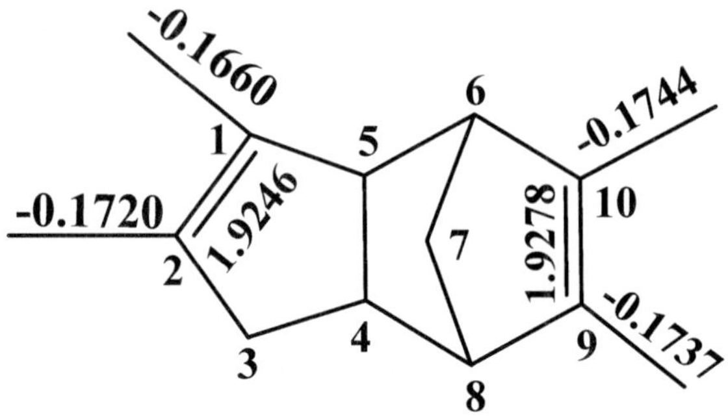

Diagram 1. The charges on the atoms and bond orders in TDC molecule.

As it's seen from diagram, the higher value of the charge observed on the carbon atoms in "1-2" and "9-10" position. This indicates that these carbon atoms are reactionable centers for electrophilic addition. Also, bond orders in "1-2" and "9-10" position are characterized with higher values. For example $P_{1,2}(\pi)$=1.9246 and $P_{9,10}(\pi)$=1.9278 (where index π showed bond orders of π components). Because $P_{9.10}(\pi)>P_{1.2}(\pi)$ it's possible to predict, that the addition is more possible with participation of $P_{9.10}(\pi)$ electrons. This conclusion does not except the addition in "1-2" position.

The hydrosilylation reaction was studied at the constant temperature, in the range 60÷90^{0}C. It was established that during the reaction in melt condition, at beginning stages of addition, three dimensional systems were obtained, which may be explained by intermolecular reactions in macro chain. For the purpose of synthesis of soluble organosiloxanes the reaction were carried out in absolute toluene solution (C=0.1017 mol/l). The course of the reaction was observed by the decrease of the amount of active ≡Si-H groups (Figure 3).

It was shown that with the increase of temperature the depth of hydrosilylation increases. At 60^{0}C the conversion of active ≡Si-H groups after 5 h is about 61% (Figure 3, curve 4), at 90^{0}C the conversion is equal to 85% (Figure 3, curve 1). In case of methylhydridesiloxane-dimethylsiloxane oligomer the conversion of active ≡Si-H groups at 90^{0}C is ~75%.

The hydrosilylation proceeds virgeously at first 90-100 min, after that the conversion of active ≡Si-H bonds increases slightly. On the bases of experimental investigations it's possible to assume that hydrosilylation mainly proceeds with obtaining of various linked branching systems, according to the following scheme 1.

Where: [(x)+(k)+(y)+(z)+(l)+(q)](p)=m≈30, n=0: 90^0 C(I) m≈53, 60^{0}C(II^1); 70^{0}C(II^2), 80^{0}C(II^3), 90^{0}C (II); [(x)+(k)+(y)+(z)+(l)+(q)](p)=m≈30, m≈33, n=23: 90^{0}C (III).

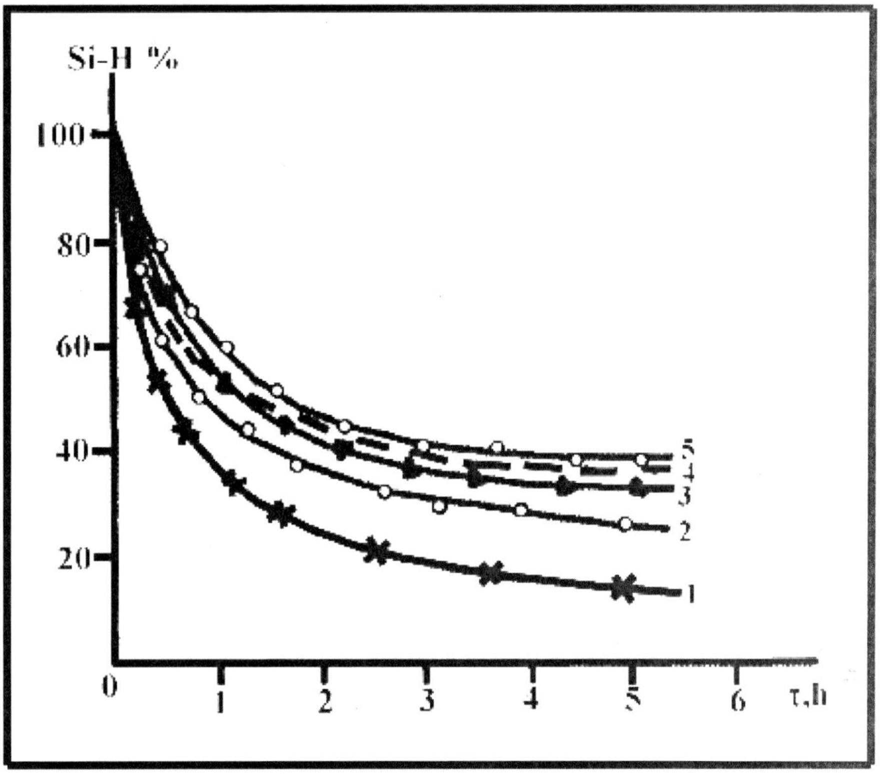

Figure 3. Dependence of changes of active Si-H% groups on the time upon of hydride addition of α,ω-bis(trimethylsiloxy)methylhydridesiloxane to tricyclidecadiene, where curve 1 is at 90^{0}C, curve 3 is at 80^{0}C, curve 4 is at 70^{0}C and curve 5 is at 60^{0}C. Curve 2 corresponds to hydride addition of α,ω-bis(trimethylsiloxy)methylhydridesiloxane-dimethylsiloxane to tricyclodecadiene at 90^{0}C.

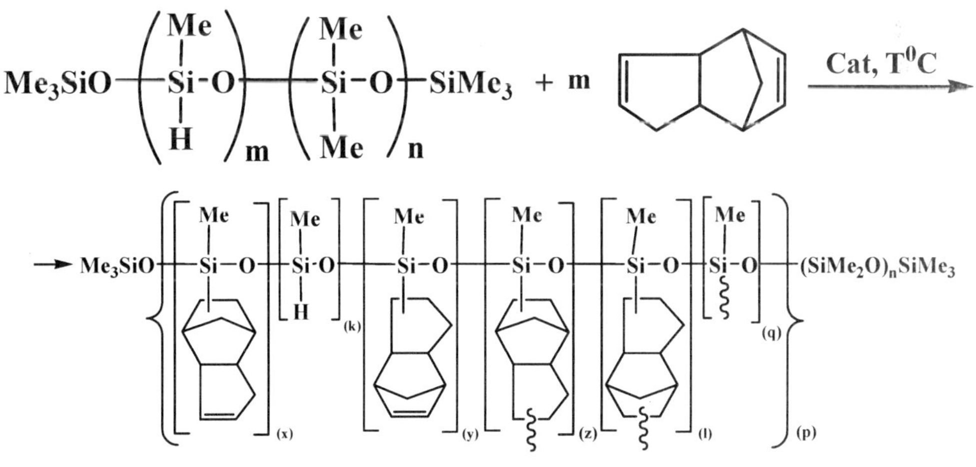

Scheme 1. Hydrosilylation reaction of α,ω-bis(trimethylsiloxy)methylhydridesiloxane to tricyclodecadiene.

It was established that with the rise of temperature cross-linked polymers are obtained. During hydrosilylation at 90^{0}C at 85% conversion of active ≡Si-H bonds about 33-35% insoluble polymers were obtained, this may be explained by the secondary interchain hydrosilylation reactions on the bases of catalyst in reaction system, according to the scheme 2:

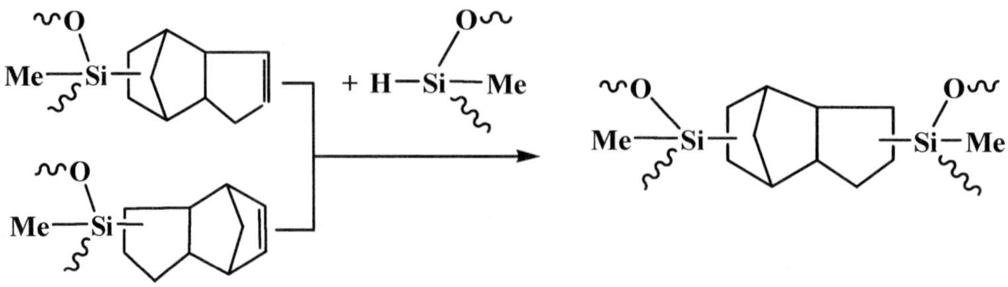

Scheme 2. Possible scheme of intermolecular hydrosilylation reaction.

Consequently in TCD, -CH=CH- bond in "1-2" position as unsaturated bond in "9-10" position is characterized with higher reaction ability reference to hydrosilylation. Therefore for reception of soluble copolymers hydrosilylation reaction stopped up to achievement 75-80 % of conversion of active ≡Si-H groups. Then the reaction mixture was connected to the vacuum (P=1÷2 mmHg) and the toluene and un-reacted TCD was removed in the temperature range 40-50^{0}C under the vacuum up to constant weight. The synthesized copolymers are transparent, opalescent viscous products well soluble in ordinary organic solvents with η=0.07÷0.13. Some physical-chemical properties of oligomers are presented in Table 2.

Table 2. Physical-chemical properties of methyl(tricyclodecenil)(hydride)siloxanes

No	Yield, %	React. Temp-re ^{0}C	η^*_{sp}	T_g, ^{0}C	$\overline{M}_\omega x10^{-3}$	P, polydisp	5%, mass losses	d_1, Å
I	85	90	0.08	-1	7.8	2.6	240	9.10
I^1	65	-	0.13	+8	82.1	21.8	300	9.33
II^1	83	60	0.07	-	-	-	-	-
II^2	75	70	0.08	+2	7.8	1.8	-	-
II^3	70	80	0.09	-	-	-	-	-
II	67	90	0.09	+6	10.3	2.3	270	9.33
III	70	70	0.11	-15	-	-	250	8.63

*In 1% solution of toluene at 25^{0}C.

In the FTIR spectra of synthesized oligomers one can observe absorption bands characteristic for -CH=CH- bonds at 1600 and 3550 cm^{-1}, the absorption bands at 2165, 1275 and 1020 cm^{-1}, characteristic for ≡Si-H, ≡Si-Me and ≡Si-O-Si≡ linear bonds accordingly.

For fully characterization of hydride addition of methylhydridesiloxane to tricyclodecadiene, by quantum-chemical semi empiric AM1 method [9], for all initial, intermediate and final products, in modeling reaction of hydrosilylation of methyldimethoxysilane to tricyclodecadiene, the heats of formations (ΔH_f), change of energy (ΔH) of the system depending on the change of distance ($R_{C\text{-}Si}$) between ≡C-Si≡ bonds, also the charges values (q) on the atoms, dipole moments (μ) and bonds or-ders (Pij) are calculated.

For the beginning the modelling reaction of hydride addition of [$Me(MeO)_2SiH$] methyldimethoxysilane to tricyclodecadiene to 9,10-direction was considered. The model reaction proceeds with formation of compound IV according to the following scheme 3.

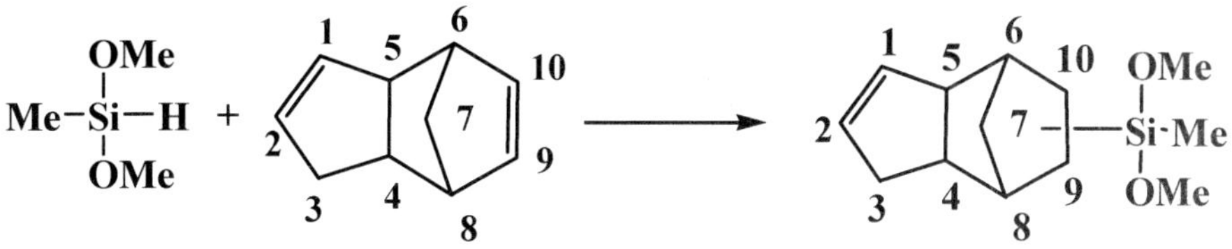

IV

Scheme 3. Modelling reaction of hydride addition of methyldimethoxysilane [$Me(MeO)_2SiH$] to tricyclodecadiene 9.10-addition.

The distances between the double bonded carbon atoms (C_{10}, C_9) and silicon atoms in tricyclodecadiene was taken about 1.0 Å more than the bond, expected in a product.

The change step of distances ($R_{C_{10}-Si}$) between atom of silicon and double bonded carbon atom (C_{10}) was ~0.05 Å. Dependence of change of energy (ΔH) of the system on the distances ($R_{C_{10}-Si}$) between silicon and carbon atoms is presented on the Figure 4.

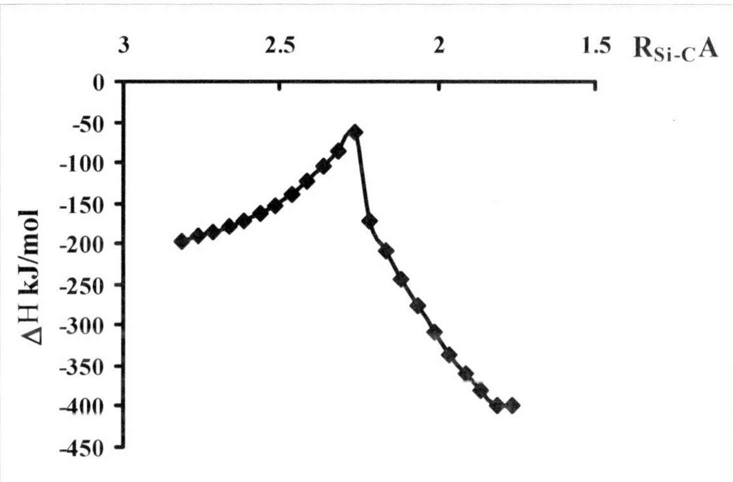

Figure 4. Dependence of changes of energy (ΔH) on distances ($R_{C_{10}-Si}$) between silicon and carbon atoms during of modelling hydrosilylation reaction of methyldimethoxysilane to tricyclodecadiene (9.10-addition).

As it is seen from Figures 4, at approximation of silicon to atom of carbon (C_{10}) up to $R_{C_{10}-Si}$=2.26 Å distances the energy of system rises. At the same time the bond order between of carbon C_{10} and C_9 atoms in a molecule of tricyclodecadiene decreases from 1.909 up to 1.699. The bond order of between hydrogen and silicon atoms also decreases (R_{Si-H}=0.922-0.775) and in modelling compound methyldimethoxysilane the process of formation of new bonds ($R_{C_{10}-Si}$=0,015-0,105 and P_{C_9-H}= 0.0006-0.063) is observed. In case of distance up to 2.21 Å the energy of system sharply decreases, the double C=C bond passes in ordinary C-C bond ($P_{C_{10}-C_9}$=0.991), and as to the bond order between silicon and carbon atom C_{10} that it reaches to 0.798. Thus the hydrogen completely eliminated from silicon (P_{Si-H}=0.013) and attach to carbon atom C_9 (P_{C_9-H}=0.929). The same results have been obtained during hydride addition of silicon and double bonded C_9 carbon atom.

With consecutive reduction of distances the energy of system monotonously decreases and with strengthening formed new bonds. The heat of formation of the product IV of hydride 9.10-addition (ΔH=-399.129 kJ/Mole) is in a good correlation with calculated heat of formation (ΔH=-399.012 kJ/Mole) of compound IV obtained by hydride addition of methyldimethoxysilane to tricyclodecadiene at 9.10-direction. Activation energy of 9.10-addition of modelling reaction is equal to E_{act}=133.21 kJ/Mole.

The 1.2-addition of modelling compound methyldimethoxysilane to tricyclodecadiene is consider-ed also. The modelling reaction of 1.2-addition proceeds with formation of compound V according to the scheme 4:

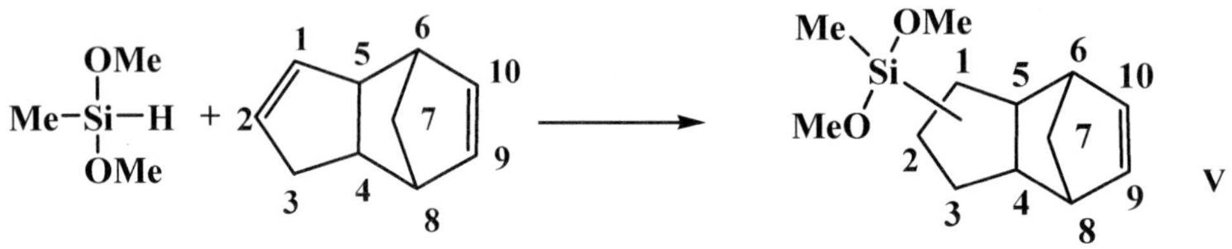

Scheme 4. Modelling reaction of hydride addition of methyldimethoxysilane [$Me(MeO)_2SiH$] to tricyclodecadiene 1.2-addition.

On the Figure 5 the dependence of change of energy (ΔH) of the system on the distances (R_{C_1-Si}) between silicon and carbon atom is presented. Similarly the above-mentioned the distance between the double bonded carbon atom (C_1) and silicon was taken about 1.0 Å more than the bond, expected in a product. The change pace of distances (R_{C_1-Si}) between atom of silicon and double bonded carbon atom (C_1) was about 0.05 Å.

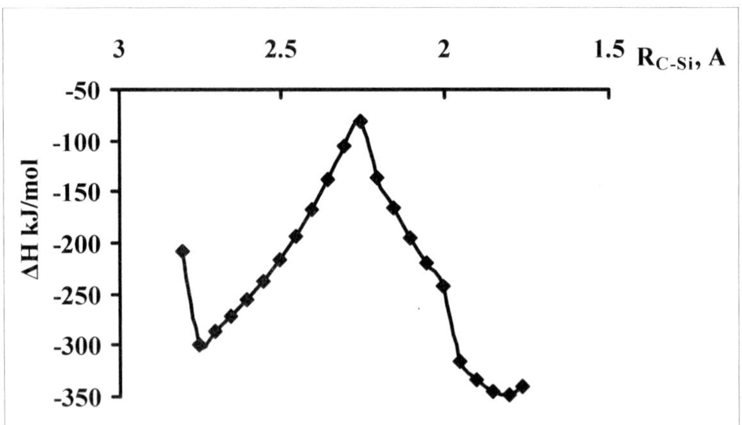

Figure 5. Dependence change of energy (ΔH) on distance (R_{C_1-Si}) between silicon and carbon atoms during modelling hydride 1.2-addition of methyldimethoxysilane to tricyc-lodecadiene.

As it is seen from Figures 8, at approximation of silicon to atom of carbon (C_1) up to $R_{C_1\text{-}Si}$=2.30 Å distances the energy of system rises. At the same time the bond order between of carbon C_1 and C_2 atoms in a molecule of tricyclodecadiene decreases from 1.901 up to 0.978. The bond order of between hydrogen and silicon atoms also decreases ($R_{Si\text{-}H}$=0.917-0.884) and in modelling compound methyldimethoxysilane the process of formation of new bonds ($R_{C_1\text{-}Si}$=0,0085-0,853 and $P_{C_2\text{-}H}$= 0,0009-0,011) is observed. In case of distance up to 2.25 Å the energy of system sharply decreases, the double C=C bond passes in ordinary C-C bond ($P_{C_1\text{-}C_2}$=0,980), and as to the bond order between silicon and carbon atom C_1 that it reaches to 0,853 Thus he hydrogen completely eliminated from silicon ($P_{Si\text{-}H}$=0,043) and attach to carbon atom C_2 ($P_{C_2\text{-}H}$=0.887). The same results have been obtained during hydride addition of silicon and double bonded C_2 carbon atom.

With consecutive reduction of distances the energy of system monotonously decreases and with strengthening formed new bonds. The heat of formation of the product V of hydride 1.2-addition (ΔH=-399.129 kJ/Mole) is in a good correlation with calculated heat of formation (ΔH=-399.012 kJ/Mole) of compound V obtained by hydride addition of methyldimethoxysilane to tricycledecadiene at 1.2-direction. Activation energy of 1.2-addition of modelling reaction is equal to E_{act}=218.67 kJ/Mole.

Thus at comparison of the Figures 4 and 5, and also energy of activation (the character of curve of change of energy and the value of bond orders) one can be concluded, that the course of modelling reaction of hydride addition of methyldimethoxysilane to tricyclodecadiene energetically is more favourable by 9,10-addition, that is in agreement with NMR spectral data.

In the Figure 6 the dependence of reverse concentration of the reactant on the time is presented. One can see that at the beginning stages the hydrosilylation is the reaction of the second order. The hydrosilylation reaction rate constants at various temperatures were calculated: K_{60^0C}= 1.051x10^{-1}, K_{70^0C}=1.405x10^{-1}, K_{80^0C}=1.944x10^{-1} mol/l·sec.

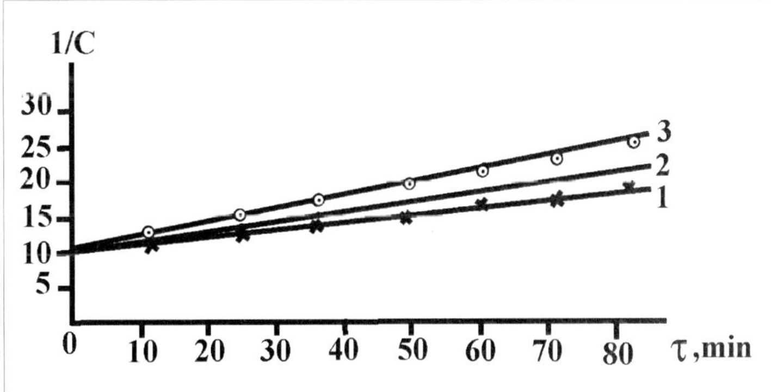

Figure 6. Dependence of reverse concentration on the time, where curve 1 is at 60^{0}C, curve 2 is at 70^{0}C and curve 3 is at 80^{0}C.

From the dependence of logarithm of the hydrosilylation reaction rate constants on the reverse temperature the activation energy of hydrosilylation reaction of methylhydridesiloxane to tricycledecadiene was calculated ~E=32.3 kJ/mol.

The synthesized oligomers were studied by GPC method. Figure 7 shows monomodal molecular weight distribution of oligomer I. $\overline{M}_{\omega}\approx7.8x10^3$ and polydispersity is equal to P≈2.6. Whence it follows, that in case of full hydrosilylation $\overline{M}_{\omega}$ does not exceed $6x10^3$. Increased molecular weight once again proved that the hydrosilylation reaction proceeds to macromolecular interchain hydride addition.

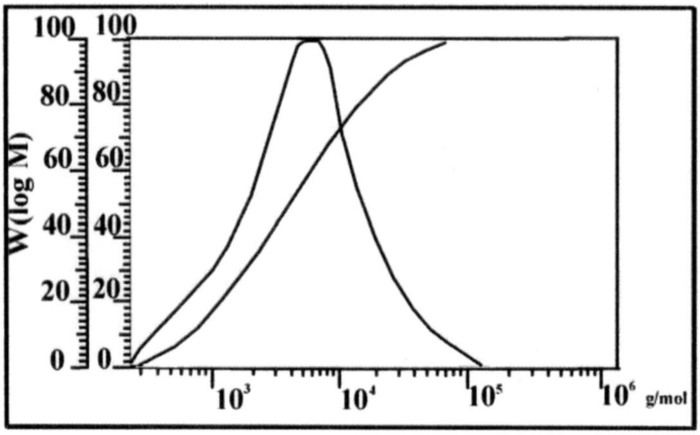

Figure 7. Gel permeation chromatographic curves of oligomer I.

The oligomer I (30% toluene solution) was heated additionally for 4-5 hours at 80-90^{0}C temperature interval, at this time the value of active ≡Si-H group's decreases from 18% up to 3%. After filtration the toluene was removed at 40-50^{0}C in vacuum (P=1÷2 mmHg) up to constant mass and transparent yellow oligomer I^1 (65%) was received, which is soluble in ordinary organic solvents.

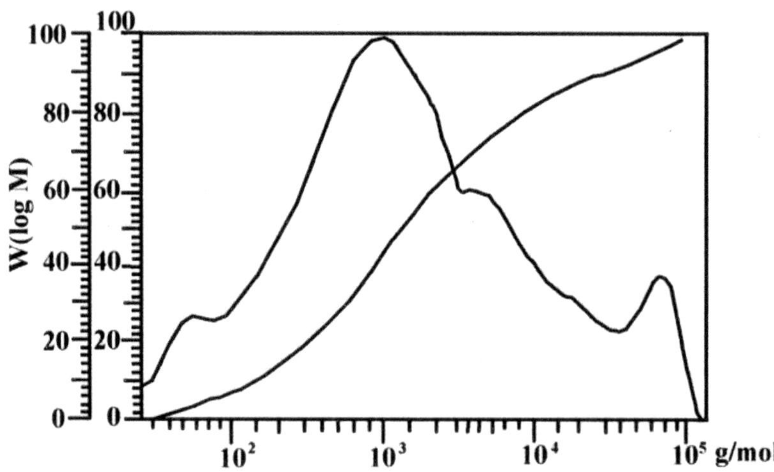

Figure 8. Gel permeation chromatographic curves of oligomer I^1.

By GPC analysis Figure 8, showed that during additional heating the macromolecular intermolecular hydride addition takes place and polydispersity oligomer I^1 with $\overline{M}_{\omega}\approx8.3x10^4$ and P≈21.8 were received.

The TGA analysis of methylsiloxane oligomers containing tricyclodecenil in the side chain was carried out and it was shown that they are characterized by higher thermal oxidative stability than trimethylsiloxy group blocked polydimethylsiloxanes. As it's seen from Figure 9, 5% mass loses for obtaining for oligomers II^2 and III is observed in the range 290-300^{0}C. The main destruction process proceeds in 350-700^{0}C the temperature range, after 700^{0}C the mass losses did not observed. The oligomer III characterize by lower thermal-oxidative stability, then oligomer II^2.

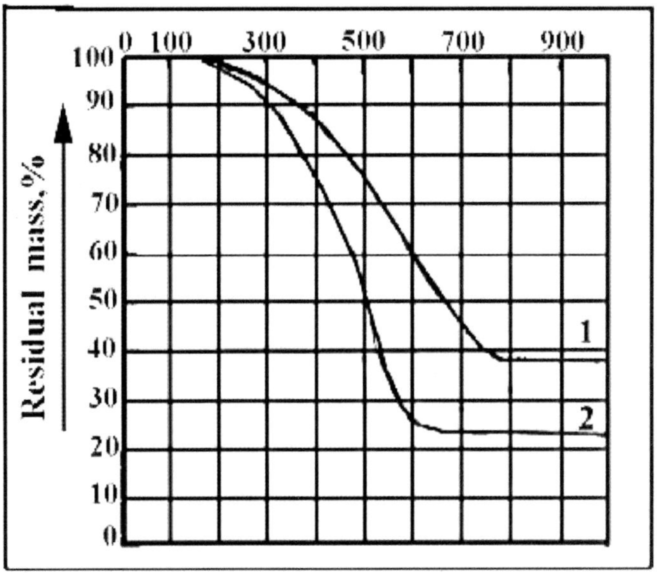

Figure 9. Thermogravimetric curves of oligomers, where curve 1 is for oligomer II^2 and curve 2 is for oligomer III.

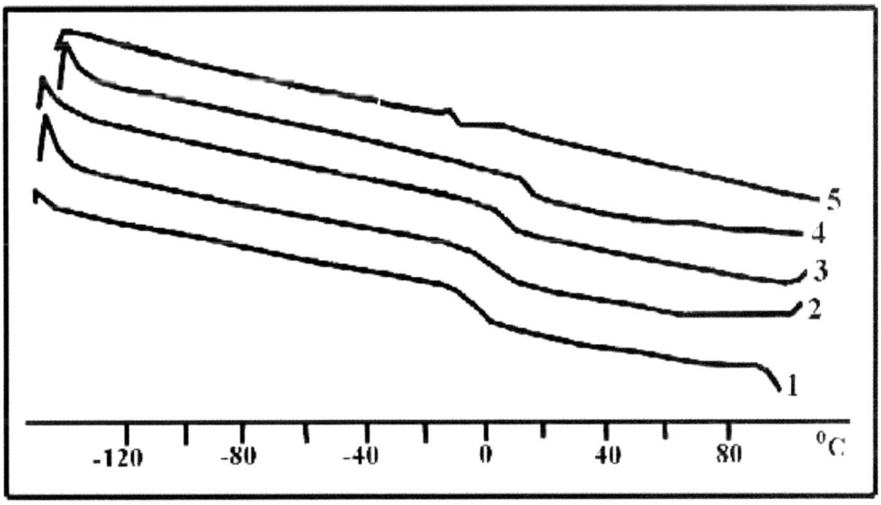

Figure 10. DSC curves of oligomers, where curve 1 is for oligomer I, curve 2 is for oligomer II^1, curve 3 is for oligomer II, curve 4 is for oligomer I^1 and curve 5 is for oligomer III.

By DSC analysis (Figure 10) it was shown that the first temperature transition for oligomer I, which corresponds to glass transition temperature (T_g) observed at ~-1^{0}C and for oligomer I^1 it observed at ~+8^{0}C. With an increase of amount of tricyclodecenil fragments in the linear siloxane chain (with an increase of the depth of hydride addition) the T_g of oligomers rises (see Figure 9, curves 1-4). For oligomer III glass transition temperature is equal to ~ -15^{0}C.

The tricyclodecenil containing methylhydridesiloxane oligomers are amorphous one phase systems. With an increase of amount of tricyclodecenil fragments in dimethylsiloxane chain the value of interchain distances d_1 slightly increases.

The synthesized tricyclodecenil and hydride containing linear methyl siloxanes are interesting products, because they contain reactionable functional groups which easily enter in intermolecular addition reaction at the expense of in system existence platinum hydrochloric acid, that gives us possibility to use such oligomer to prepare siliconorganic rubbers.

REFERENCES

[1] Mukbaniani, O.V.; Khananashvili, L.M. *Intern. J. Polymer. Materials,* 1994, 27, 31.

[2] Aneli, G.N.; Koberidze, Kh.E.; Mukbaniani, O.V.;.Karchkhadze, M.G.; Khananashvili, L.M. *Russian Polymer News*, 1998, 3, 7.

[3] Galashina, N.M. *Vysokomol. Soed.*, 1994, 36A, 640.

[4] Chauhan, B.P.S.; Rathore, J.S.; Glloxhani, N. *Appl. Organometal. Chem.* 2005, 19, 542.

[5] McGrath, M. P.; Sall, E. D.; Tremontm, S. *J. Chem. Rev.* 1995, 95, 381.

[6] Marciniec, B. In Comprehensive Handbook on Hydrosilylation, Pergamon Press, Oxford, 1992.

[7] Aneli, J.N.; Mukbaniani, O.V.; Karchkhadze, M.G.; Koberidze, Kh.E.; Tkeshelashvili, R.Sh.; Khananashvili, L.M. *Georgian Patent,* N1348101-96.

[8] Nogaideli, A.I.; Khananashvili, L.M.; Nakaidze, L.I.; Chogovadze, T.V.; Tskhovrebashvili, V.S.; Volkova, R.V.; Nogaideli, G.A. Dokl. Akad. Nauk GSSR, 1978, 92, 341.

[9] Magomedov, G.K.I.; Shkolnik, O.V.; Khananashvili, L.M.; Nakaidze, L.I.; Katsitadze, M.G. *Zhurn. Obshch. Khimii,* 1988, 59, 104.

In: New Topics in Monomer and Polymer Research
Editors: A. D'Amore and G. Zaikov, pp. 23-33
ISBN: 1-60021-436-3

Chapter 3

THE PHYSICAL AGEING PROCESS IN POLYMERS: INFLUENCE OF SOME LIKE-NANOSTRUCTURES STABLE LOCAL CONFORMATIONS ON THE MECHANISMS

N. Guarrotxena

Instituto de Ciencia y Tecnología de Polímeros (ICTP), Consejo Superior de Investigaciones Científicas (CSIC). C/ Juan de la Cierva,3. 28006, Madrid, Spain. Phone: 34-91 562 29 00; Fax: 34-91 564 48 53. E-mail: nekane@ictp.csic.es

ABSTRACT

Two series of polymers having different tacticity-dependent stereochemical microstructures have been submitted to controlled ageing. One series consists of three PVC samples prepared at distinct temperatures and of the same samples after substitution reaction to various degrees. And one series consists of three commercial samples of PMMA. In both cases a straight correlation between enthalpy loss during ageing and the content of mmr tetrads termini of isotactic sequences of at least one heptad in length is found, especially with those mmr structures that take the GTTG⁻TT conformation. In the case of PVC that is more quantitavely evidenced through the steroselective removal of this conformation as the result of substitution reaction. Anyway, the results shed light on the mechanisms at nanometric level of the ageing process in polymers.

Keywords: physical ageing; local chain conformation; nanostructures; poly (vinyl chloride) (PVC); poly (methyl methacrylate) (PMMA).

INTRODUCTION

Physical ageing is an important phenomenon which introduces some limitations into polymer applications. It is generally manifested by change in a property of the polymer as a function of storage of application time at constant temperature and under neither stress nor influence from any other external conditions. A huge amount of work on this subject has been reported[1-8]. The most common situation exhibiting physical ageing occurs when a material is cooled from higher to lower temperatures than the glass transition temperature. Significant changes in magnitudes such as specific volume or enthalpy are observed. It is generally admitted that physical ageing phenomenon obeys the tendency of amorphous polymers below Tg, i.e. under non-equilibrium state, towards the equilibrium state which is achieved at temperatures above Tg. Needless to say that physical ageing is of great significance scientifically and industrially. The main difficulty to explain this complex phenomenon resides in the lack of a full knowledge of the interaction between on the one hand the fundamental thermodynamic quantities such as enthalpy, entropy and volume, that are known to change during the physical ageing, and on the other hand, the physical properties.

In addition, the relation between the polymer structure, at macroscopic or microscopic level, and the chain rearrangement changes that are common with ageing, are far from a complete knowledge.

The technological significance emerges from the fact that most of the processing procedures for thermoplastics involve drastic drops of temperature and this causes the ageing to initiate. Similarly in most of the technical uses, the polymer materials are subjected to changes in temperature, what can also give rise to physical ageing.

Physical ageing has been widely surveyed in the literature[9]. It may be there observed how detailed and exhaustive are the theoretical models that have been proposed[9]. What is interesting for the purpose of this paper is that most of the structural studies regarded free volume and free volume distribution as main concepts to explain the physical ageing. In the same way the cooperativity between distinct regions, in the chain rearrangements that occur during a physical ageing, has been emphasized[10]. Nevertheless the question arises as to whether these concepts are applicable without distinction, to the phenomena occurring on macroscopic and microscopic scale. Moreover, the works mentioned above give no indication as to whether free volume or molecular rearrangements are connected with chemical structural aspects or not, irrespective of their macroscopic or microscopic nature.

Such an observation is rather general to properties pÛher than physical ageing. Actually it has been often shown that there exists some relationship between a given property (mechanical, thermal relaxation, electrical, etc) and the type of polymer family. But the way by which the overall chemical structure intervenes in the specific quantity that is responsible for the property considered, is frequently unknown. The problem is much more evident as chemical structures e.g. isolated units, unusual chemical order, local stereochemical variations, etc are considered. Thus most of the theoretical treatments refer to quantities like chain segments, free volume, morphology, etc that are common to all or any polymers.

With respect to chemical microstructures a series of works to be mentioned is the one developed in our laboratory, specially with PVC and PP as reference materials[11-15]. The microstructures considered are the following: a) the isotactic and syndiotactic sequences, that relate to the extent to which a polymer exhibits a non-Bernoullian tacticity distribution; b) the mmr-based and the rrrm-based local structures which occur necessarily whenever an isotactic sequence or a syndiotactic sequence breaks off respectively; and c) the atactic parts and the pure mrmr moieties. On the other hand the mmr-based structures can be of two distinct types namely mmrm and the mmrr structures. The former one relates always to isotactic sequences whereas the latter can exist at the end of an isotactic or a syndiotactic sequence indistinctly.

The ratio of mmrm to mmrr will obviously increase with increasing non-Bernoullian isotacticity.

The content of the above all microstructures may by either determined accurately by ^{13}C-NMR spectroscopy or estimated qualitatively from experimental measurements[14].

The overall microstructures have been shown to be determining in chemical modification and degradation of PVC and in properties such as molecular interactions with solvents, esters and polyesters[16]; glass transition[17,18] (Tg), β relaxation and antiplasticization[19]; dielectric relaxations[20,21] and electrical properties (nature, distribution and stabilization of space charges)[22-25]. The correlations obtained have been recently extended to physical ageing of PVC and PMMA. The results, likewise for all the properties above mentioned, inferred not only a straight relation between physical ageing and stereochemical microstructure, but also between the latter and the local quantities that are liable to initiate physical ageing as generally suggested by the literature (free volume, rotational chain mobility with or without coupling, etc).

It must be finally emphasized that the above microstructures are all nanometrical in size. The results are then applicable to nanoscience, a matter of high interest nowadays.

This article deals with approaches to physical ageing knowledge as inferred from recently found correlations with the stereochemical microstructures mentioned above.

EXPERIMENTAL

Materials

Poly(vinylchloride). The PVC samples used were an-additive-free industrial PVC prepared in bulk polymerization at temperatures of 68°C (sample X), 60°C (sample Y) and 45°C (sample Z). The samples were characterized as described elsewhere[11,15]. Cyclohexanone (CH) was purified as reported earlier[11]. Dioctyl phthalate (DOP) was used as received. Samples X, Y and Z were modified by nucleophilic substitution reaction with sodium benzenethiolate (NaBT) in CH and in the melt in the presence of DOP moieties following the methods described in previous papers[11,15]. The tacticity of modified samples was measured by means of ^{13}C NMR decoupled spectra as described elsewhere[11,15].

Poly(methyl methacrylate). PMMA (MAX, MAY and MAZ) samples from Atochem, were purified using tetrahydrofuran (THF, Scharlau) as solvent and water as precipitating agent, washed in methanol and dried under vacuum at 40°C for 48 hours. Tetrahydrofuran was distilled under nitrogen with aluminium lithium hydride (Aldrich) to remove peroxides immediately before use.

Characterization of Samples

The Tacticity of All Samples Was Measured as Follows

a. poly(vinylchloride).The tacticities of both the starting (samples X, Y and Z) and the modified polymers were measured by means of ^{13}C NMR decoupled spectra obtained at 80°C on an Varian UNITY-500 instrument, operating at 125 MHz using 1,4-dioxane-d_8 as solvent under conditions described previously[11,15]. The resonances used were those of methine carbons of the backbone ranging from 57 to 61 ppm. The 40000-50000 scans gave a very satisfactory signal-to-noise and the respective peak intensities were measured from the integrated areas, as calculated by means of an electronic integrator.

b. poly(methylmethacrylate). ^{1}H-NMR spectra were recorded on a Varian UNITY-500 spectrometer operating at 499.88 MHz in $CDCl_3$ as solvent at 50°C. The parameters of 8000Hz spectral width and a 1.9s pulse repetition rate were used. The delay time was set to 0.5s. The spectra were obtaining after accumulating 64 scans with a sample concentration of 10 wt% solutions. The relative peak intensities were measured from the integrated peak areas, which were calculated with an electronic integrator. The content of the different triads were determined by the electronic integration of the α-methyl signals which appear at (1.3-1.1ppm), at (1.1-0.9ppm) and at (0.9-0.7ppm) corresponding to isotactic (mm), heterotactic (mr) and syndiotactic (rr) triads, respectively. It shows the main signals related to stereomicrostructure.

Calorimetry Measurements

a. *poly(vinylchloride).* The physical ageing of either unmodified PVC (samples X, Y and Z) (Table 1) and their appropriate modified polymers with NaBT at 25°C in purified CH and in the melt were studied by DSC using a Mettler TA 3000. The polymer samples, in amounts ranging from 10 to 11 mg, were encapsulated in aluminium pans. They were aged according to the following thermal treatment: (a) annealing of the sample at 130°C for 10 min; (b) cooling down the sample to the ageing temperature (10°C below Tg) at 40°C min^{-1} and ageing for 60 min; (c) quenching the sample down to 30°C; and (d) performance of a heating DSC run from 30°C to 130°C at 10°C min^{-1}. In order to obtain a measure of the enthalpy relaxation, a reference DSC curve was obtained repeating the thermal treatment without ageing.

Table 1. ^{13}C-NMR data of PVC samples

Sample	Mn	Triads			Tetrads	Pentads			Bernoullian Probabilities			h[a]
		mm	mr	rr	mmr	mmmm	mmmr	rmmr	P_{mm}[b]	P_{rm}[b]	P_{mm}[b] − P_{rm}[b]	
X	42000	20.1	49.65	30.25	22.65	3.97	9.61	6.52	0.447	0.451	(0)0.004	0.8
Y	53500	18.89	49.38	31.72	20.31	4.21	9.06	5.63	0.443	0.438	0.005	0.6
Z	64000	18.66	49.36	31.99	20.68	3.77	9.11	5.79	0.421	0.431	(-)0.01	0.5

[a] Content of liabile structures[21]

[b] Calculated probabilities (see text)

b. polymethylmethacrylate. About 11mg of PMMA sample was placed in an aluminium pan and put on a hot plate (170°C, N_2) where it was maintained for 10min to erase previous thermal history; it was then cooled to 50°C at a fast rate (40°C/min).

Subsequently, the samples were moved into a Mettler TA 300 instrument and annealed at Tg-10°C, for the following times: 30, 45, 60, 120, 240 min (MAX); 7, 15, 30, 60 min (MAY) and 15, 30, 45, 60, 120 min (MAZ). Finally they were quenched to 50°C and performanced of a heating DSC run from 50 to 170°C at 10°C min^{-1}. In order to obtain a measure of the enthalpy relaxation, a reference DSC curve was obtained by repeating the thermal treatment without ageing.

RESULTS AND DISCUSSION

The stereochemistry-based nanostructures in PVC and PMMA. The NMR spectroscopies allow one to state and to measure the above quoted microstructures[26]. ^{13}C-NMR and ^{1}H-NMR spectroscopies were used to analyse the sorts of mmr- and rrm- based stereochemical sequences and their surroundings for one series of PVC samples and one series of PMMA samples, of different tacticities. Basically, the spectra allow to detemine the content of isotactic (mm), heterotactic (mr and rm) and syndiotactic (rr) triads. In fact, what is directly measured on the spectra are the integrated areas of signals belonging to the coresponding pentads. In the case of the isotactic pentad of PVC the spectra resolution is good enough for the content of the three sorts of pentad, mmmm, mmmr and rmmr to be detemined accurately., From the triad content it is possible to calculate on the one hand the content of isotactic (m) heterotactic (h) and syndiotactic (r) dyads and on the other hand, the probability that one placement r follows one m palacement or one r placement, P_{mr} and P_{rr} respectively. The corresponding equations have been given previously[26]. Obviously, if Bernoullian statistics applies then P_{mr} =P_{rr}. Therefore, P_{mr} – P_{rr} is an acurate measure of the sample departure from Bernoullian behaviour (Note that the same is valid for P_{mm} and P_{rm} denoting the probability that one m placement follows one m or one r placement, respectively). In addition in case of Bernoullian behaviour the probability of occurrence of any repeating sequence may be obtained through universal relations[26]. The results so obtained for the samples of this work are presented in Tables 1 and 2. As can be seen PVC samples, prior to modification (Table 1) are all Bernoullian. Thus there appear to be not so different in tacticity arrangement. Nevertheless the whole isotacticity clearly decreases from sample X to sample Z. As widely argued[26] this means that the isotactis sequenes of at least one heptad in length decreases in the same order and so does the content of the conformation GTTG$^-$TT of the tetrad mmr which occcurs necessarily whenever an isotactic sequence breaks off, relative to the other likely conformation, the GTGTTT[26]. The content of GTTG$^-$TT conformation may be determined by means of substitution reaction at as low temperature as –15°C. The values so obtained are denoted by *h* in Table 1. The implications of the substitution raction to the tacticity arrangement are commented below.

With reference to the PMMA tacticity it appeas clear that samples MAX and MAY are totally isotactic particularly sample MAX, while sample MAZ is non Bernoullian syndiotactic. The overall isotactic content decreases from sample MAX to sample MAY. Therefore what has been said in the precedent paragraphs applies to PMMA. Sample MAZ will exibit more syndiotactic sequence content. On the other hand, the content of rrm tetrad, which occurs whenever one syndiotactic sequence breaks off, will be grater. More details concerning PMMMA tacticity are given below.

The role of some like-nanostructures stable local conformation on the physical ageing mechanism. The results obtained for the PVC samples (Table 1) are given in Figure 1. This is typical behaviour of any polymer. An endothermic peak appears as a consequence of the lost enthalpy occurring during ageing[1,17]. This quantity may be determined from the DSC curves as the difference in the areas between the ageing temperature and 130°C, the temperature at which equilibrium has been attained in both cases.

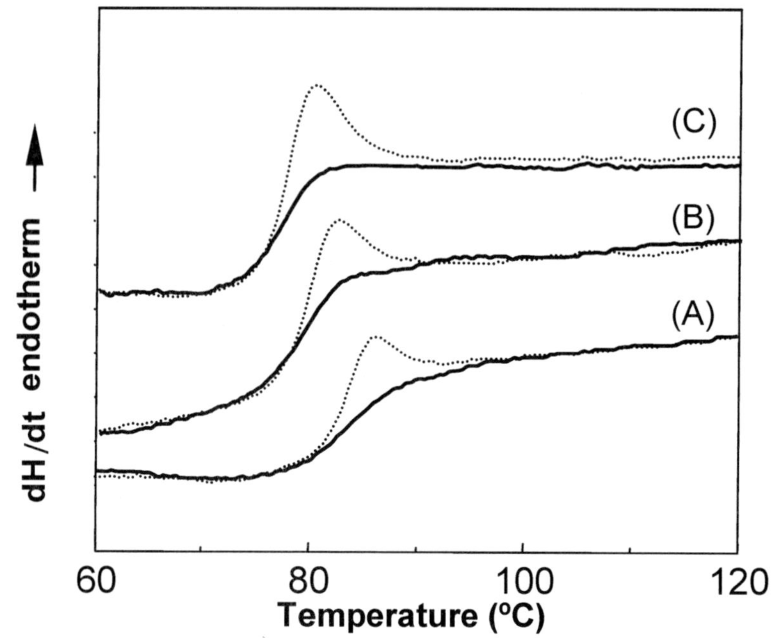

Figure 1. Endothermic peak in DSC curves of unaged (—) and aged (…) PVC (sample X) at different degrees of substitution: (A) 0%; (B) 6.5%; (C) 12%.

The evolution of the thermal energy absorbed from the difference between the areas under the two curves for every sample vs. degree of substitution reaction achieved in solution and in the melt state respectively[11,15] is displayed in Figures 2 and 3. As can be seen, the absorbed heat passes through a minimum at a definite degree of substitution, which agrees with the content of GTTG⁻ in the polymer (Table 1, *h* values) for all PVC samples. Actually, the higher the content of mmr, irrespective of its conformation, involved in the substitution reaction, the higher the absorbed heat upon ageing. Then it increases up to a conversion of roughly 7% and afterwards the trend is towards levelling off.

As extensively argued in earlier work[11-13] such a reaction involves: i) the depletion of the mmr structures that take the GTTG⁻TT conformation; ii) the progressive disappearance of mmr under GTGT conformation that is associated with isotactic sequences of a least one heptad; and iii) the decrease of the remaining mmr and the rrmr structures terminal to syndiotactic sequences, simultaneously. These well defined processes have been demonstrated to occur, respectively, during the ranges of conversion 0.5-0.8%, 1-7% and 7-12%. Afterwards it is the rrmr structure that reacts principally. Therefore, the microstructure of PVC samples agree with the occurrence of processes (i), (ii) and (iii).

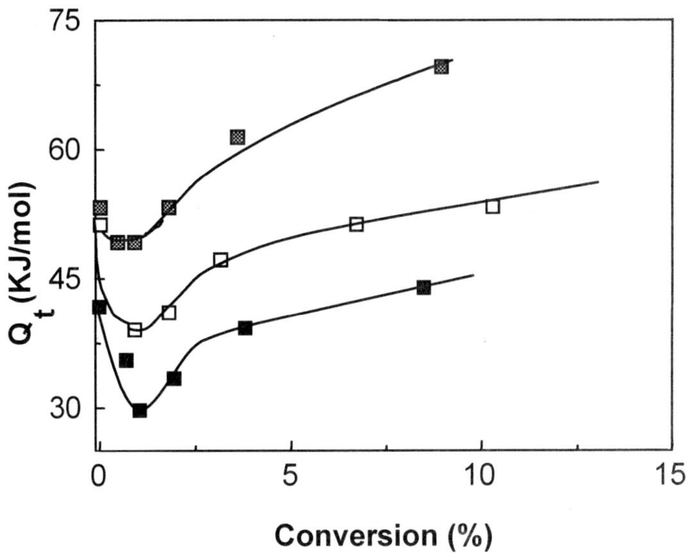

Figure 2. Dependence of the area of the endothermic peak (Q_t) on the degree of substitution of PVC in CH solution: (■) sample X; (□) sample Y; (▩) sample Z.

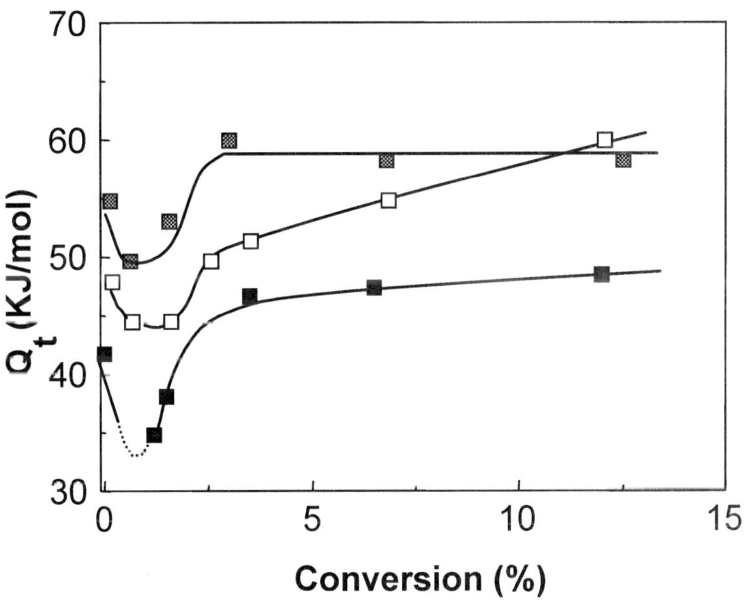

Figure 3. Dependence of the area of the endothermic peak (Q_t) on the degree of substitution of PVC in melt state: (■) sample X; (□) sample Y; (▩) sample Z.

Such a procedure should allow one to correlate the changes in physical ageing, if any, with the tacticity dependent microstructure fitting into some fundamental achievements on the microstructure/physical property relationships as attained in earlier work[19]. The occurrence of a minimum is a quite novel result in that , unlike in the above quoted attempt[19], it allows one to obtain an original insight into the mechanisms of physical ageing.

Now, as indicated before, at conversions between 0.5-0.8% and 7%, the mmr tetrads under GTGTTT conformation associated with isotactic sequences of at least one heptad length start reacting preferably[11,15] producing little lost enthalpy relative to the GTTG⁻TT conformation during the ageing experiment.

After about 7% conversion, no isotactic sequences equal or higher than one heptad are available and the reaction proceeds through either short isotactic sequences or the mr heterotactic triad in mmr[11,15].

The higher effect of GTTG⁻TT conformation for an equal modification quantity, as shown in Figure 2, is consistent with its higher free volume and local mobility as comparing to what happens with the remaining reactive structures[14].

The results obtained for the same reaction in the melt state are displayed in Figure 3. As can be seen these results give support to the straight relation between GTTG⁻TT conformation and ageing. The minimum is more profound and occurs at higher conversions as the content of GTTG⁻TT of the sample increases (Z<Y<X) (Table 2). Overall, some interesting conclusions were drawn from the substitution reaction in the melt state despite the severe reaction conditions (shear stress, temperature, amount of plasticizer and load): 1) this reaction conditions only affect the kinetics and, somewhat, the nucleophile efficiency and 2) they influence neither on the evolution of the tactic triads nor on the molecular weight with conversion[28]. It was actually found that the substitution reaction proceeds most frequently by GTTG⁻TT conformation.

Table 2. ^{1}H NMR data of the PMMA samples

		Trialds			Bernoullian probabilities		
Sample	M_n	mm[a]	mr[a]	rr[a]	P_{mm}[b]	P_{rm}[b]	P_{mm}[b] - P_{rm}[b]
MAX	45503	8.56	41.81	49.63	0.291	0.296	(-)0.005
MAY	43131	12.8	39.92	47.28	0.391	0.297	0.094
MAZ	44868	20.97	36.5	42.53	0.535	0.3	0.235

[a] Direct measurements from NMR spectra
[b] Calculated probabilities (see text)

From the above results it follows that physical ageing of PVC strongly depends on the tacticity dependent microstructure, in particular, the local configurations located at the end of isotactic sequences.

Figure 4 plots the DSC curves for samples MAX, MAY and MAZ of poly (methyl methacrylate) prior to and after ageing for the indicated annealing times in the Experimental section. These DSC curves obtained in aged samples show, as expected, an endothermic peak at Tg as a result of the structural relaxation process that the material undergoes. The height and area of this endothermic peak change with the ageing condition of the material, increasing the area of its maxima as the annealing proceeds as a result of the lost enthalpy occurring during ageing[1,27].

This enthalpy relaxation calculated according to the substraction of the DSC curves prior to and after ageing appears in Figure 5. As suggested áj a mere glance at Figure 5 curves, the slope of them is greater as the ageing period-time increases up to a given value. Then the trend is towards a plateau in the three polymers indicating a diminution on the free volume and consequently in the local chain molecular motions. Interestingly, the curve amplitude, i.e. enthalpy lost or relaxation subsequent to ageing, decreases in the order MAX>MAY>MAZ.

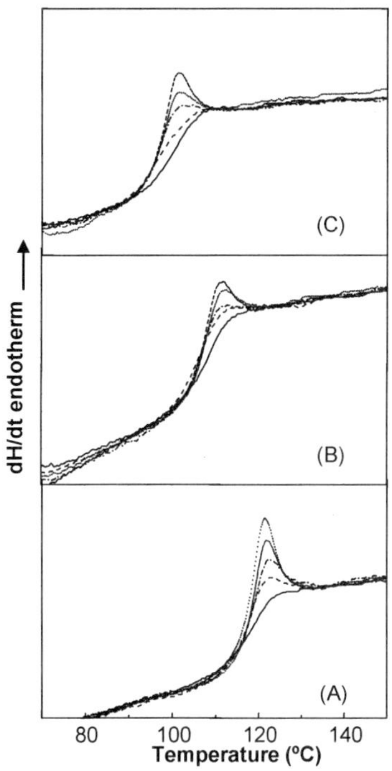

Figure 4. Endothermic peak in DSC curves of PMMA annealed at Tg-10°C at different times: [(A, sample MAX: 30; 60; 120; 240 min); (B, sample MAY: 7; 15; 30; 60 min); (C, sample MAZ: 15; 30; 60; 120 min)] and the corresponding unaged DSC curves (—).

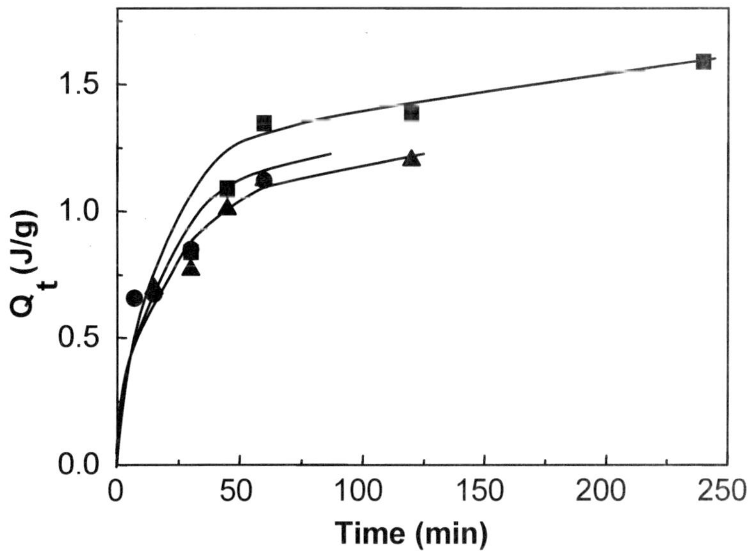

Figure 5. Dependence of the area of the endothermic peak (Q_t) on the ageing time (min) of PMMA: (■) sample MAX; (●) sample MAY; (▲) sample MAZ.

Comparing these behaviours and those previous reported for PVC it is evident that the occurrence of a plateau is a clear indication that some well defined structures, no matter their type determine the physical ageing process. In addition, the differences between the slope of curves (Figure 5) is consistent with the different contents of mmr structures associated with isotactic sequences of the polymers.

This is what was proved for PVC where the stereospecific mechanisms of substitution[11,14,15,29] involves a controlled disappearance of one mmr under GTTG⁻TT or GTGTTT conformation by exchanging it for a sequence with a GTTTT conformation of very little chain mobility. By comparing this correlation and that obtained for PMMA (Figure 5) a variance with expectations were found. In effect, a characteristic minimum attributable to the content of mmr associated with isotactic sequences taking the GTTG⁻TT conformation (Figures 2 and 3) was now absent in the left part of the curves in Figure 5. This behaviour is consistent with the subsequent hypothesis that the GTTG⁻TT conformation is less favoured in PMMA. A valuable explanation of these results lies in the fact that mmr is more rigid in PMMA case disturbing that the GTTG⁻TT can be taken, unlike in PVC. It may then be inferred that the mmr, whenever related to sufficiently long isotactic sequences, is liable to enhance the departure from enthalpy-temperature equilibrium of the material. The extent to which some of these structures adopt GTTG⁻TT conformation depends of the type of polymer. It turns out to be insignificant in PMMA relative to PVC.

Anyway the results obtained herein allow a new insight into the mechanism of the processes involved in the physical ageing, in that the latter appears to relate the ability of local configurations like mmr to experience conformational changes between two conformations of distinct free volume and mobility that can occur in that configuration. And it may be inferred that mmr structure termini of isotactic sequences equal or higher than one heptad is a major driving-force for the physical ageing process likewise for a number of physical properties[19].

ACKNOWLEDGEMENTS

I am grateful to the Comisón Interministerial de Ciencia y Tecnología (C.I.C.Y.T.) for financial support (MAT2002-00250) of this research.

REFERENCES

[1] Struik L.C.E., in *Physical Ageing in Amorphous Polymers and Other Materials*. Elsevier Science, Amsterdam, The Netherlands, 1978.

[2] Kovacs, AJ., *Adv. Polym. Sci.*, 1963, 3, 394.

[3] Hutchinson, J.M. and Kovacs, A.J., *J Polym. Sci. Polym. Phys. Ed*, 1976, 14, 1975.

[4] Petrie, S.E.B., *J. Polym. Sci. Part A-2*, 1972, 10, 279.

[5] Foltz, C.R. and McKinney, P.V., *J. Appl. Polym. Sci.*, 1969, 13, 2235.

[6] Kovacs, A.J., Akloins, J.J., Hutchinson, J.M: and Ramos, A.R., *J. Polym. Sci. Polym. Phys. Ed.*, 1979, 17, 1097.

[7] O'Reilly, J.M., *J. Appl. Phys.*, 1979, 50, 6083.

[8] Chen, H.S. and :Wang, T.T., *J. Appl. Phys.*, 1981, 52, 5898.

[9] Hutchinson, J.M., *Prog. Polym. Sci.*, 1995, 20, 703.

[10] Ngai, K.L., *Comments on Solid State Physics* 1980, 9, 141.

[11] Guarrotxena N, Martínez G and Millán J, *J Polym Sci Polym Chem* 1996, 34, 2387.

[12] Guarrotxena N, Martínez G and Millán J, *J Polym Sci Polym Chem* 1996, 34, 2563.

[13] Guarrotxena, N., Martínez G and Millán J, *Acta Polymerica* 1999, 50, 180.

[14] Guarrotxena N, Schue F, Collet A and Millán J, *Polym Int* ., 2003, 52, 420.

[15] Guarrotxena, N., Martínez, G. and Millán, J., *Eur. Polym. J.*, 1997, 33, 1473.

[16] Tiemblo, P., Martínez, G. And Millán, J., *J Polym Sci Polym Chem*1995, 33, 1243.

[17] Guarrotxena N, Martínez G and Millán J, *Polymer* 1997, 38, 1857.

[18] Guarrotxena N, Martínez G and Millán J, *Polymer* 2000, 41, 3331.

[19] Guarrotxena, N., Tiemblo, P., Martínez, G., Gómez-Elvira, J.M: and Millán, J., *Eur. Polym. J.*, 1998, 34, 833.

[20] Guarrotxena N, del Val, J.J., Elícegui, A. and Millán J, *J Polym Sci Polym Phys* 2004, 42, 2337.

[21] Elícegui, A., del Val, J.J., Millán J. And Mijangos, C., *J of Non-Crystalline Solids* 1998, 263,623.

[22] Guarrotxena, N., Vella, N., Toureille, A. and Millán J, *Macromol Chem Phys.*, 1997, 198, 457.

[23] Guarrotxena, N., Vella, N., Toureille, A. and Millán J, *Polymer* 1997, 38, 4253.

[24] Guarrotxena, N., Contreras, J., Toureille, A. and Millán J, *Polymer* 1999, 40, 2639.

[25] Guarrotxena, N., Contreras, J., Toureille, A. and Millán J, *Polym Bull* 1998, 41, 355.

[26] Tonelli AE, *NMR spectroscopy and polymer microstructure: The conformational connection.* ATandT Bell Laboratories, VCH Publishers, Inc., New York, 1990.

[27] Petrie, S.E.B., J. Macromol. Sci. Phys., 1976, B12, 225.

[28] Mijangos, C., Gómez-Elvira, J.M:, Martínez, G. and Millán, J., *J. Appl. Polym. Sci.*, 1989, 38, 1685.

[29] Martínez, G. and Millán, J., *J. Polym. Sci. Polym. Chem.*, 2004, 42, 6052.

In: New Topics in Monomer and Polymer Research
Editors: A. D'Amore and G. Zaikov, pp. 35-48
ISBN: 1-60021-436-3

Chapter 4

Generalization of Data Concerning the Coal Swelling in Organic Solvents and Their Extraction Using the Linear Multiparametric Equations

***L. I. Bazylyak*[1], *D. V. Bryk*[2], *R. G. Makitra*[2], *R. Ye. Prystansky*[1], *G. E. Zaikov*[3]**

[1] Physical Chemistry of Combustible Minerals Department
L. M. Lytvynenko Institute of Physical–Organic Chemistry and Carbon Chemistry
National Academy of Sciences of Ukraine
3a Naukova Str., 79053, Lviv, Ukraine; *e–mail: vfh@org.lviv.net*
[2] Institute of Geology and Geochemistry of Combustible Minerals
National Academy of Sciences of Ukraine
3a Naukova Str., 79053, Lviv, Ukraine; *e–mail: vfh@org.lviv.net*
[3] N. Emanuel Institute of Biochemical Physics; Russian Academy of Sciences
4 Kosygin Str., 117977, Moscow, Russia; *e–mail: chembio@sky.chph.ras.ru*

Abstract

Approaches to the consideration of a coal swelling process, which were used up to now and based on the theory of regular solutions, do not give the possibility to generalize quantitatively the experimental data. Adequate relation between the physical–chemical properties of the solvents and the degree of a coal swelling in them can be obtained only with the use of linear multiparametric equations which take into account the effects of the all processes proceeding in the system; besides, the basicity and a molar volume of the liquids are determinative. Such approach is effective at the generalization of data concerning to extraction of a coal.

The action of organic solvents on natural polymers combustible minerals (coal and brown coal or peat) is intensively studied for a long time due to following reasons. Firstly, this is one of the successful method of studying the structure of combustible materials and the second is their technological application for obtaining of a so–called montan–wax or low–molecular liquid extracts which can be transformed into synthetic liquid fuel due to hydration process. Moreover, an interaction of a coal with the solvents is a basis of the coals liquation processes and coals transformation into liquid fuel.

The first stage of an interaction in the system "*coal–solvent*" is a swelling, or an increasing in the volume, as a result of introducing the molecules of a liquid into interstices and directly into the structure of a coal. Depending on the solvent nature and the coal nature, the volume of a coal can increase even in some times, and, respectively the weight growth of investigated sample can be achieved to 100 and more percentages.

A review of early works concerning to a coal swelling is represented in [1, 2]. Let us notify, that else in the fifties the coals swelling process was considered as the first degree of their extraction and was connected with physical–chemical interaction of these coals with the solvents [3]. Such process was explained from two points of view, namely: either this process was coursed by adsorption of the liquid into interstices or such process was connected with the change of the cohesion energy of solid and liquid phases of a system. *Sanada* and co–authors [4 – 6] were taken into account, that the coal is natural three–dimensional polymer and in accordance with the *Flory–Huggins's* theory a change of free energy at the coal swelling is conventional sum of the energies of mixing the polymer and solvent and first of all is determined by the disparity of solubility parameters of both components accordingly to *Hildebrand* parameter δ:

$$\Delta G = [ln(1 - \varnothing_2) + \chi\varnothing_2^2 + \varnothing_2], \quad (1)$$

where $\varnothing_2$ is volumetrical part of netting structure (polymer) in swollen system; and parameter χ indicating the interaction of a polymer with the solvent which is equal to

$$\chi = \frac{\beta + (\delta_1 - \delta_2)^2 V_1}{RT} \quad (2)$$

also contains the empirical term β, correction factor, which takes into account the number of branching in the structure of polymer; δ_1 and δ_2 are *Hildebrand's* parameters of solubility for the solvent and polymer and are equal to $[(\Delta H_{evapor} - RT) / V_m]^{1/2}$.

After insignificant transformations the *Flory–Renner's* equation can be obtained. Such equation helps to calculate the sizes of polymer link between the cross bonds M_c

$$M_c = \frac{\rho_2 V_1 \varnothing_2^{1/3}}{[-ln(1-\varnothing_2) - \chi\varnothing_2^2 - \varnothing_2]}, \quad (3)$$

where ρ_2 is the density of a polymer into solution; V_1 is molar volume of the solvent.

Coefficient χ should be determined empirically for every solvent and, of course, from the concentration dependence of the osmotic pressure and with taken into account a series of assumptions. In the work *Sanada* [5] and subsequent works of other authors, the swelling degree in the volumetric parts Q is represented as a function from the δ of solvents. These data in most cases form the parabolic, "belfry"–like curve with a maximum for the solvents,

δ_2 of which accordingly to a theory of regular solutions is equal to or is near to δ_1 of polymer (coal). In reality, it was maintained already in the work [5] that for the coal only approximated dependencies are obtained – a number of experimental data concerning to Q are visibly take one's leaved from the generalizing curve. It was determined in the work [4] at the extraction by solvents in the Soxhlet's apparatus of vitrain from Yubary field (the content of carbon consists of 85,2 %), that the maximal yields of an extract are observed at their molar volume about 10 $cm^3/mole$ (ethylendiamine, dimethylformamide, cyclohexanone 24 – 26 %, acetophenone 35,6 %, pyridine 33,2 %). Such results were explained by the influence of the value of cohesion energy. There are a plenty of exclusions, for example for butanole $V_m = 9{,}5$, but the yield of the extract in such a case is only 0,8 %. The explanation of this deviation as a result of the solvent association caused by the presence of hydrogen bond seems unconvincing since under the experiments conditions (the extraction in the Soxhlet's apparatus, and that is under boiling temperature) the association will be insignificant. It was discovered in the work [6], that the value M_c for japanese coal with the carbon content less than 80 % is unreal low – only 10 *(!)*, next this value is sharply increased and is achieved the maximum $M_c = 175$ at 85 ^0C and after that is decreased. Authors starting from following positions explained this fact: firstly, experimental determinations were carried out in pyridine, in which specific interactions can take place and, the second this deviation can be explained by the mistakes at the determination of χ coefficient.

The same approach was discussed in the work *Kirov* and co–authors [7] in detail on example of swelling (and extraction) for three kinds of bituminous Australian coal. These authors confirmed the main observations of *Sanada* – the swelling degree Q increases from ~ 1,4 in hydrocarbons to ~ 2 in pyridine (δ – 11,0) and again decreases to ~ 1,5 in alcohols. Calculated on this basis value δ of coal increases droningly with increasing the content of carbon from 70 % till 87 % and in a case of more metamorphized coal is sharply decreased again. Data concerning to the extraction of *Greta* coal are evidence of maximal yield of extract (more than 20 %) under it treatment with ethylendiamine and dimethylformamide (δ – 11,5), however, authors admit a fact that this is a consequence of specific interactions, since in alcohol from the δ of the same order the yield of the extract is only 1 – 2 %.

Authors concluded, that although the swelling degree is not directly connected with molecular characteristics of absorbed liquids, however determining factor is their parameter of solubility in spite of the fact that at detailed consideration of the dependencies $Q = f(\delta)$ (or $f(\delta^2)$) there are a number of deviations (as same as in the work [5]) from the ideal curve for many solvents. It is necessary to notify that although it is hard to estimate the verisimilitude of determined in such a way molecular weights of structural links of a coal between the points of cross bonds, however, in a case of synthetic polymers in a same way determined masses of links visible don't agree with the values obtained in accordance with others methods.

In spite of the indicated lacks, the described above approach is applied to later works concerning to coal swelling and results interpretation. It is necessary to distinguish a plenty of investigations devoted to swelling studies of coal № 6 from Illinois State (standard in USA coal for the carbon–chemical investigations) [8 – 10]. General conclusions are in good agreement with the results of the works [5, 7]. Comparison of swelling degree for different coal in some solvents depending on the content of carbon has been done in the work [11]. Similar investigation for Siberia Kansk–Achynsk coal was carried out in the works [12, 13]. In both cases, as same as in a work [7], it was proved the dependence of swelling degree of coal on the carbon content in it.

That is why logically to assume the possibility of specific interactions also during the swelling process, since the values of parameters of the coal solubility δ^2, which are determining accordingly to the *Flory–Renner's* equation are differed. It depends on fact if the data for all solvents are taking into account in calculations or such calculations are performed with the exclusion of results for solvents able to be as acceptors of hydrogen bonds (amines, ketones). Different results have been obtained also under application of other methods for calculations, especially of the *Van–Krevelen's* method [14].

It is notified in the work [15], that the swelling of some coal does not agree with the thesis of regular solutions theory; that is why, it is not allowed to calculate the parameter χ for them. Authors explain this fact by the presence of oxygen atoms in the investigated coal. But also the molecular weight of separate sections (clusters) between the points of crossing for methylated or acetylated samples of this coal is equal only to 300 – 600 in accordance with the calculations (that is unreal).

It is necessary to notify, that the critical analysis of the *Flory* theory application for the determination of molecular mass and the crossing density of the coal structure has been done in the *Painter's* works [16]. Authors assert, that the possible formation of hydrogen bonds between the hydroxy groups of low–metamorphized coal has an important role here; that is why, even a lot of empirical amendments introduction into calculations leads to obtaining the understated values of molecular masses of clusters.

Taking into account the above–mentioned lacks authors concluded that the theory of regular solutions is insufficient for adequate description of the coal swelling process (and also for the extraction process) in different solvents since such theory does not take into account the possible specific solvation of active structures of coal and first of all its heteroatoms [17] especially by formation of hydrogen bonds. With the aim of taking into account the possible acid–base interactions it was proposed by *Marzec* and co–authors [18, 19] to determine the swelling degree as a function of donor number of *DN* solvents or as a function of their donor and acceptor numbers disparity accordingly to *Gutmann*. However, corresponding analysis of data concerning to swelling the shlessian bituminous coal showed the following: although between the Q and *DN* is visible symbasis, however the deviations from the straight line is less than for the function $Q = f(\delta)$; but, at the same time it is complicated to confirm about the quantitative description of the process. The same conclusion about only qualitative character of such dependence has been done by authors [13] on example of swelling the brown Kansk–Achynsk coal and some kinds of Donbas coal.

Above–mentioned facts and disagreements lead to the conclusions [20] that the sorption of solvents by coal is very complicated process, which covers also the changes under the action of solvent into the coal structure and other possible non–equilibrium phenomena. That is why, application for a coal the theories developed for the description of thermodynamically equilibrium process of swelling the simple synthetic polymers is unwarranted first of all due to neglect the existing chemical (specific) solvation interactions. As it was confirmed in many investigations, the swelling *Flory–Huggins's* model based on the theory of regular solutions is not sufficiently consistent with the real experimental data. It is caused by a range of simplifying assumptions putted into the base of this model and, first of all, the presence of full isoentalpic mixing (solution) of two phases that is in disagreement with the reality – even in a case of the polymers which do not contain the donor–acceptor groups into the structure a swelling and solution processes are accompanied with a great enthalpy effect; it is know, that even non–specific solvation is often accompanied by the changes of free energy and enthalpy of the system. And isoentalpy will be not remained in a case of the possible donor–acceptor (acid–base) interaction, which is often observed in a case of synthetic polymers with the content of heteroatoms (polyurethane, nitryle rubbers) and is observed in a case of coal as a result of the presence in it such groups as *–OH*, *–COOH*, tertiary atom of nitrogen and *ect.* Calculations on the basis of the theory of regular solutions for the coal swelling have mostly unsatisfactory generalizing and predicted ability.

The principle of the linearity of free energies *(LFE)* is applied in chemistry of solutions over 30 years for quantitative description of the solvents influence on the behavior of dissolved substances (spectral characteristics, constants of the reaction rate). In accordance with this principle general change of free energy of the system consists of the separate inter–independent terms and first of all consists of non–specific and specific solvation and also needed energy for the formation of cavity in the structure of liquid phase with the aim of allocation the exterior molecule introducing there. And only full sum of these all possible energetic effects gives the final (equilibrium) energy of the system [21]:

$$\Delta G = \sum \Delta g_i \tag{4}$$

With taken into account, that the constants of the reaction rates are determined via the equilibrium constants of the activated reactive complex formation, and the last in part depend on the solvation processes, it was proposed by *Koppell* and *Palm* [22] the following equation in order to determine the influence of medium properties on the reaction rates of processes proceeding in it:

$$\lg K = a_0 + \frac{a_1(n^2-1)}{(n^2+2)} + \frac{a_2(\varepsilon-1)}{(2\varepsilon+1)} + a_3 B + a_4 E_T \tag{5}$$

This equation takes into account the influence of the polarization $f(n^2)$ and polarity $f(\varepsilon)$ of the solvents determining their ability to non–specific solvation and also their basicities B and electrophilicity accordingly to Reichardt E_T characterizing their ability to introduce into acid–base interactions (specific solvation). Appropriateness of this equation for the generalization of experimental data of the dependencies of reactions rates (and also spectral characteristics of dissolved substances) on physical–chemical characteristics of the solvents has been proved by a number of hundred examples.

We proposed to expand the *Koppell–Palm's* equation by fifth term which takes into account the density of the energy of solvents cohesion proportional to the squared *Hildebrand's* solubility parameter δ^2. Due to this fact necessary energy for the formation of cavity for the allocation of the molecule introducing into liquid phase is taking into account:

$$\lg K = a_0 + a_1 f(n^2) + a_2 f(\varepsilon) + a_3 B + a_4 E_T + a_5 \delta^2 \tag{6}$$

Modified equation was turned out effective for the determination of the solvents influence on the equilibrium of such processes as solubility in different media of gases or solids, the distribution of substances between two phases, resembling equilibrium processes. So, it will be logically to try to use the equation (6) for the swelling processes. As a matter of fact, it was turned out, that with the use of this equation it is possible to determine the quantitative connection between the properties of the solvents and equilibrium swelling degree of a number of polymers, and also of a coal [23 – 25]. In order to achieve the satisfactorily high values of the coefficients of multiple correlations *R*, it is necessary to exclude from the calculations the data for some quantity (3 – 5) of the solvents. It is hard to explain it. Besides, it was not quite clear the model of the interactions into the system. In a case when the solvation processes are energetically advantageous ($\Delta G < 0$) and that is why promote to the swelling process, that is to the solvent penetration into the structure of polymer, then the role of δ^2 factor is remained not clear. Such factor characterizes the energy needed for the cavity formation into the structure of the liquid; at the same time, unlike to the evaporation process, under the swelling of substances into liquid the following process takes place: liquid solvent penetrates into the structure of solid polymeric phase mostly as the whole.

At the beginning of ninetieth the works of *Aminabhavi* are appeared [26]. These worked were concerned the polymeric membranes swelling into organic solvents and to diffusion rate *D* of the liquids into their structure in which these values were considered as dependencies from the molar volume V_M of the liquids. Generalizations obtained in [26] are rather unsatisfactory – approximately linear dependencies lgQ or lgD on V_M are observed only in the homologic ranges or in the case of similar solvents. But approach must be considered as logical: it is clear, that in a case of bigger sizes of introducing molecule, the last with difficulty will be penetrated into the structure of polymer including the adsorbent interstice. Low generalizing ability of the dependencies presented in the work [26] can be explained by fact that they do not take into account the solvation effects, which promote to liquids penetration. That is why the equation (6) has been expanded by additional term, which takes into account the influence of molar volume of the solvents:

$$lg\,Q = a_0 + a_1 f(n^2) + a_2 f(\varepsilon) + a_3 B + a_4 E_T + a_5 \delta^2 + a_6 V_M \qquad (7)$$

Such equation under the stipulation that *Q* is represented not in the volumetric parts accordingly to the *Flory–Huggins's* model but in accordance with the interpretation of equilibrium processes in the chemical thermodynamics as a moles of the solvent absorbed by one gram or by one cm^3 of polymer was turn out effective under the generalization of data for swelling degree of different synthetic polymers, for example, polyethylene, in different organic solvents depending on their physical–chemical parameters [27]. That's why, it was necessary to check the possibility of application the equation (7) for the generalization of data concerning to coal swelling since this equation takes into account the all important possible energetic effects caused by possible donor–acceptor interaction of active groups of the coal with non–inert solvents including the formation of hydrogen bonds; the effects of non–specific coal solvation with solvents which are caused by a presence in it the cyclic aromatic structures as a result of which the visible influence of the ability of some solvents for the polarization can be expected; and also endothermic effects as a result of steric complications of the solvents penetration *(V_M)* and destruction of the liquid phase structure *(δ^2)*.

Data concerning to a swelling of the most popular coal (namely, coal *Illinois № 6*) have been taken by us as a main object of our investigations. This coal is the standard object for the carbon–chemical investigations in USA. These data were already analyzed earlier in works

[23 – 25] with the aim of their generalization accordingly to equation (6), but obtained results were unsatisfactory. Evidently this was caused by two factors: *i)* the influence of the molecules sizes of the solvents penetrating into the coal structure (their molar volume) was not taken into account in these works; *ii)* analyzed starting values of the swelling degree Q were given accordingly to original works [9 – 11] in *ml* (sometimes in *g*) of the solvent absorbed by 1 *ml* or 1 *g* of coal since the *Flory–Huggins's* model has been used by authors (this model uses the volumes of two liquids which are mutually mixed). If to consider the coal swelling process (and generally polymers swelling processes) as thermodynamically equilibrium process then the free energy change at the penetration and at the absorption of a solvent and, respectively, the equilibrium constant of this process, it is preferable to determine the quantity of connected solvent in molar units.

In the presented paper we have checked the efficiency of the above–mentioned factors considering studying the swelling process in organic solvents. Illinois coals are low–metamorphized, bituminous and contain 20 – 31 % of volatile substances, are characterized by ash content 8 – 12 % and sulphur content 4 – 7,8 %. Investigated in the work [9] sample was characterized by following composition: C 79,8; H 5,11; N 1,8; $S_{org.}$ 2,0 and O 11,2 %. The samples were pulverized and extracted from the soluble components with pyridine; after vacuum–drying they were saturated by solvent's steams till their full saturation at room temperatures in the closed vessels. Authors give the ratio of the weights for swelled samples respectively to the starting W. Generalization of these data for 10 solvents in accordance with the fifth–parameter equation (6) leads to the expression with unsatisfactory low value of correlation coefficient R = 0,81 [23]. Exclusion from the consideration the most uncoordinated data for dioxane gives the possibility to obtain the fifth–parameter equation with low, but acceptable degree of connection R = 0,941. At the same time, consideration of the molar volume factor and the change of weight parts on the molar ones essentially improve the correlation – for all 10 studied solvents R = 0,940, and after the exclusion the data concerning to cyclohexane for the rest 9 solvents we obtain the equation with high connection degree R = 0,996 [28].

There are only two decisive parameters – *first is basicity* which assists to swelling process and *the second is molar volume*, which opposites to this process; taking into account the needed energy and the negotiation of the cohesion forces have only insignificant influence and the exclusion of this parameter from the calculations practically does not worsen the equation

$$lg\,Q = -1{,}96 + (0{,}665 \pm 0{,}074)10^{-3}B - (4{,}12 \pm 0{,}58)10^{-3}V_M\,;$$
$$R = 0{,}984 \text{ and } \quad S = 0{,}030 \qquad (8)$$

Obtained equations have greater predicted ability comparatively to fifth–parameter equation obtained in the work [23], which does not take into account the factor of molar volume (for nine solvents R is equal to 0,940).

The influence of the factor of molar volume is confirmed by fact that between lgQ and V_M the neatly marked symbasis is observed, namely: with increasing of V_M the value lgQ is decreased. The action of other decisive factor – solvents basicity – is opposite, in other words with the basicity increasing the symbate increasing of lgQ is observed. Evidently, this is caused by the specific solvation of acid centers, which are in the macromolecule of coal and, first of all, of hydroxy groups. Their presence can be assumed taking into account a great number of the oxygen in the Illinois coal. Comparison of these two oppositely directed dependencies leads to the conclusion, that they are mutually compensated. Although these dependencies are only symbate, but the algebraic sum of the influence of these two factors is practically linearly connected with the respective values *lgQ* [29]. The third influenced factor is the density of the cohesion energy, which is proportional to needed energy for separation of absorbed molecules from the structure of liquid phase; this factor respectively also decreases the swelling value. However, the influence of this value is insignificant; this fact is confirmed by negligible decreasing of the *Q* value at its exclusion. The possible processes of non–specific and electrophilic solvation practically do not impact on the value *Q*.

In the work [11] authors also have been studied the swelling process of the coal Illinois № 6 in the liquid phase. Swelling degree *S* has been studied by volumetrically as the ratio of volumes of swelling sample to the starting one. Unlike to [8], it was investigated the process in a range of amines including the primary ones, able to the formation of hydrogen bonds and also alcohols. At the generalization of these data in accordance with the fifth–parameter equation without taking into account of V_M for the all 17 solvents it was obtained the equation with the low value $R = 0{,}861$; but at the use of the sixth parameter equation (7) and after the exclusion of data for isopropanole and dimethylaniline we achieve of high correlation

$$lg\,Q = -2{,}91 + (0{,}454 \pm 1{,}40)f(n^2) + (5{,}73 \pm 1{,}22)f(\varepsilon) + (1{,}43 \pm 0{,}37)10^{-3}B - (29{,}9 \pm 22{,}0)10^{-3}E_T - (0{,}722 \pm 0{,}947)\delta^2 - (6{,}52 \pm 4{,}54)10^{-3}V_M$$

$$N = 15 \qquad R = 0{,}980 \qquad S = 0{,}160 \qquad (9)$$

and after the exclusion of insignificant factors

$$lg\,Q = -4{,}34 + (3{,}42 \pm 0{,}69)f(\varepsilon) + (2{,}02 \pm 0{,}27)10^{-3}B - (0{,}86 \pm 2{,}55)10^{-3}V_M$$

$$R = 0{,}968 \text{ and } \quad S = 0{,}172 \qquad (10)$$

In this case the basicity and the molar volume of the solvents are decisive factors, the influence of which is oppositely directed. An appearance of the polarity as significant factor is connected with the specific selection of high polar solvents (alcohols, amines).

Calculated in accordance with the equation (10) values $lgQ_{calc.}$ and their deviation from the experimental values are represented in *Table* 1.

Accordingly to [9] the swelling process of the Illinois coal № 6 has been carried out principally under other conditions, namely: the samples were previously extracted with pyridine, dried coal was standed till the full saturation with vapors at 100 ^0C in closed metallic ampoules (with the exception of phenol, investigating temperature of which is 182 ^0C). Authors presented the results of investigations as the ratio of swelling *W* (in percentages) that is the ratio of weights of swelling sample after 1 hour to the dried sample. These data have been previously generalized in the work [24]. Low value *R* for the all 12 solvents equal to 0,876 after exclusion from the consideration data concerning to the phenol and tetrahydrophurane is increased till 0,972. Essentially better results were obtained with taken into account the molar volume factor.

The data concerning to W taken from [10] and calculated on their basis swelling values in moles Q and lgQ are presented in *Table* 2; the generalization of these data in accordance with the sixth parameter equation (7) leads to higher degree of relationship R = 0,909, and the exclusion from the consideration of one solvent (butylamine) gives the possibility to obtain the equation with satisfactory degree of relationship R = 0,974; an additional exclusion of dimethylformamide gives the equation (11) with R = 0,991.

$$\lg Q = -2{,}61 + (3{,}50 \pm 0{,}82) f(n^2) + (2{,}30 \pm 0{,}46) f(\varepsilon) - (0{,}33 \pm 0{,}14)10^{-3} B - (2{,}37 \pm 6{,}8)10^{-3} E_T + (0{,}70 \pm 0{,}24)10^{-3} \delta^2 - (1{,}5 \pm 2{,}1)10^{-3} V_M$$

$$N = 10, \qquad R = 0{,}991 \quad \text{and} \quad S = 0{,}055 \qquad (11)$$

and after the exclusion of insignificant factors

$$\lg Q = -2{,}51 + (2{,}66 \pm 1{,}20) f(n^2) + (1{,}80 \pm 0{,}60) f(\varepsilon) - (7{,}4 \pm 5{,}0)10^{-3} E_T - (2{,}8 \pm 2{,}9)10^{-3} V_M$$

$$R = 0{,}964 \quad \text{and} \quad S = 0{,}086 \qquad (12)$$

With the molar volume of the solvents increasing the coal swelling degree is decreased; the same is an effect of the ability to electrophilic solvation. Unlike to both previous cases, the positive influence of the solvents basicity (namely their ability to form the donor–acceptor bonds with acid groups of the coal) here is insignificant evidently as a consequence of especial influence of the conditions of experiment carrying out. Under higher temperatures the hydrogen bonds are easy decomposed. At the same time, the possible positive influence of the factors of non–specific solvation $f(n^2)$ and $f(\varepsilon)$ is observed. Calculated values lgQ and their discrepancy with the experiment ΔlgQ are presented for the comparison in *Table* 2.

Table 1. Experimental [10] and calculated in accordance with the equation (10) values of swelling degree of the coal Illinois № 6

№	Solvent	Experiments			Calculations	
		S	$Q_m 10^3$	lgQ_m	lgQ_m	ΔlgQ_m
1	2–Picoline	2,76	17,84	–1,749	–1,794	0,045
2	Pyridine	2,75	21,72	–1,663	–1,808	0,145
3	Butylamine	2,64	16,57	–1,781	–1,941	0,160
4	Propylamine	2,45	17,62	–1,754	–1,655	–0,099
5	Aniline	1,99	10,86	–1,964	–2,205	0,241
6	2–Hexanone	1,98	8,120	–2,090	–2,312	0,221
7	Methylaniline	1,44	4,052	–2,392	–2,037	–0,355
8	Propanol	1,36	4,820	–2,317	–2,240	–0,077
9	Ethanol	1,34	5,824	–2,235	–2,194	–0,041
10	Butanole	1,34	3,715	–2,430	–2,242	–0,188
11	Methanol	1,23	5,690	–2,245	–2,204	–0,041
12	Dimethylaniline*	1,10	0,789	–3,103	—	—
13	Isopropanole*	1,06	0,784	–3,106	—	—
14	Toluene	1,06	0,562	–3,250	–3,311	0,061
15	p–Xylene	1,06	0,487	–3,312	–3,326	0,013
16	m–Xylene	1,05	0,407	–3,390	–3,293	–0,097
17	Benzene	1,04	0,447	–3,350	–3,362	0,012

Note: *data excluded from the calculations

Both the equilibrium swelling degree and the kinetics of this process depend on the character of the solvent. In the work [10] it has been studied the swelling rate of the coal Illinois № 6 by volumetrically in different solvents; on the starting stages it is ordered to the pseudo–first order reactions kinetics. It helped to determine the respective constants rate of the process, which are presented in *Table* 3. In the work [25] we have generalized these data for 24 solvents with the use of fifth–parameter equation (6). For the all maximal sequence of the data the value of correlation multiple coefficient *R* was very low and equal to 0,694 and only after the exclusion from the calculation the data for five solvents (that is practically 20 %) it could obtain the satisfactory value of *R* = 0,957. Additional taking into account the influence of molar volume, that is transition to sixth parameter equation, gives the possibility to obtain the expression with *R* = 0,883. And in order to obtain the satisfactory correlation it was enough to exclude from the calculations data for only two solvents, namely 2–hexanone (methylbutyl ketone) and triethylamine

$$lg\,k = 2{,}20 - (2{,}55 \pm 3{,}84) f(n^2) + (1{,}08 \pm 4{,}26) f(\varepsilon) + (4{,}17 \pm 1{,}12) 10^{-3} B - (71{,}7 \pm 62{,}5) 10^{-3} E_T - (0{,}92 \pm 2{,}45)\delta^2 - (42{,}8 \pm 9{,}1) 10^{-3} V_M$$

$$N = 22, \quad R = 0{,}959 \quad \text{and} \quad S = 0{,}448 \qquad (13)$$

Table 2. Experimental [9] and calculated in accordance with the equation (12) values of "swelling ratio" of soluble part of coal for the coal Illinois № 6

№	Solvent	Experiments			Calculations	
		W	$Q_m 10^3$	lgQ_m	lgQ_m	ΔlgQ_m
1	Dimethylformamide	6,2	60,54	–1,218	—	—
2	N–Methylpirrolodone	5,7	37,17	–1,430	–1,503	0,073
3	Dimethylsulphoxide	5,5	49,19	–1,308	–1,407	0,099
4	Ethylendiamine	4,6	33,24	–1,478	–1,466	–0,012
5	Aniline	4,6	34,13	–1,467	–1,513	0,047
6	Butylamine*	3,8	29,82	–1,525	—	—
7	Pyridine	3,7	28,67	–1,543	–1,475	–0,068
8	Phenol	3,4	22,49	–1,648	–1,512	–0,136
9	Pipyridine	3,0	19,26	–1,543	–1,475	–0,068
10	Tetrahydrofuran	2,8	22,97	–1,639	–1,615	–0,024
11	Toluene	2,6	16,65	–1,779	–1,874	0,095
12	Hexane	1,6	6,902	–2,161	–2,134	–0,027

Note: *data excluded from the calculations

The equation terms characterizing the influence of non–specific solvation and also cohesion energy have a great standard deviations which are more than the absolute values of the coefficients and that is why are evidently insignificant. Checking the value *R* decreasing at the exclusion of these terms confirmed this assumption and helped to obtain the equation with lesser quantity of significant terms. This equation also adequately characterizes the influence of the solvents properties on the rate of their penetration into the coal structure; besides, the decisive factor in this case as same as in a case of swelling value is the influence of molar volume of the solvents, increasing of which leads to the process rate decreasing.

$$lg\,k = 1{,}12 + (4{,}85 \pm 0{,}52)10^{-3}B - (66{,}0 \pm 19{,}5)10^{-3}E_T - (42{,}1 \pm 6{,}2)10^{-3}V_M$$

$R = 0{,}957$ and $S = 0{,}418$ (14)

Significant factor as same as in a case of swelling degree is the solvents basicity. With the solvents basicity increasing, the process rate is also increased. The less essential is a role the solvents ability to electrophilic solvation; although this factor increases the process rate but it exclusion from the consideration decreases *R* till 0,928. The value *lgQ* calculated in accordance with the equation (14) is represented in *Table* 3.

Decisive role of the V_M factor during the adsorption process of the solvents by coal is in agreement with the determined in the work [**25**] proportionality for the alcohols between *lgk* and steric factor E_s of the *Hammet–Taft's* equation.

Table 3. Experimental [10] and calculated accordingly to equation (14) values of the logarithms of the constants rate of the coals Illinois № 6 swelling

№	Solvent	Experiments		Calculations	
		$k10^5$, s^{-1}	lgk	lgk	Δlgk
1	Propylamine	1167,0	–1,933	–1,679	–0,254
2	Butylamine	614,0	–2,212	–2,974	0,763
3	Pyridine	316,7	–2,499	–2,655	0,155
4	2–Picoline	126,7	–2,897	–3,124	0,226
5	2–Hexanone*	125,0	–2,903	—	—
6	Methanol	53,30	–3,273	–3,183	–0,091
7	Ethanol	21,30	–3,672	–3,624	–0,048
8	Aniline	20,0	–3,699	–3,962	0,263
9	Propanol	10,90	–3,963	–4,290	0,328
10	Butanole	3,84	–4,416	–4,926	0,511
11	Isopropanole	2,54	–4,595	–4,152	–0,443
12	Methylaniline	2,47	–4,607	–4,063	–0,545
13	Butanole–2	1,32	–4,879	–4,694	–0,185
14	Toluene	0,90	–5,046	–5,334	0,288
15	Isobutanol	0,833	–5,079	–4,859	0,220
16	Dimethylaniline	0,59	–5,229	–4,578	–0,651
17	Benzene	0,45	–5,347	–4,675	–0,671
18	Pentanol	0,376	–5,425	–5,510	0,085
19	p–Xylene	0,375	–5,426	–5,924	0,498
20	m– Xylene	0,225	–5,648	–5,919	0,271
21	o– Xylene	0,118	–5,928	–5,891	–0,037
22	Ethyl benzene	0,113	–5,947	–6,032	0,086
23	Cumene	0,009	–7,046	–6,716	–0,330
24	Triethylamine*	0,00038	–8,420	—	—

Note: *data excluded from the calculations

So, the swelling characteristics of the Illinois coal are determined by total influence of molar volume of liquids and their ability to specific solvation. The same conclusion has been done by authors of the works [30, 31] explaining the adsorption growing by increasing the donor number of the solvents via the formation of hydrogen bond by *OH*–groups of coal. But these authors have not done respective quantitative generalization giving the possibility on the basis of the linearity of free energies principle adequately to connect the properties of the liquids with their ability to interact with a coal; it was confirmed that the approaches based on the theory of regular solutions equitable only at the consideration of the swelling process in the "inert" (so–called low–basic) solvents, mainly of low–polarity.

Correctness of the sixth parameter equation (7) and its simplified forms for the generalization of the swelling data was proved for other coals including the Donbas coal [32] at the parameters B and V_M. If to apply the equation (7) to the coal extraction data, then the factor of molar volume V_M is insignificant, and the connection between quantities of extracted substance (in *g/mole* of the solvent) and physical–chemical characteristics can be satisfactorily described by fifth parameter equation (6) or by its simplified forms; in this case possible acid–base interaction is the decisive factor, that is factor B [33 – 35]. This confirmation is in good agreement with the above–said: bigger molecules harder introduce into the coal structure and after equilibrium state their size does not play the role. Let us notify, that the same approach has the positive results at the data generalization concerning to the solubility of the synthetic low–molecular coal analogous – diphenylolpropane – in 20 solvents.

This approach is also applicable for the generalization of data concerning to the coal extraction under sub–critical conditions, but the role of the specific solvation is also insignificant, evidently as a result of its suppression at high temperatures.

So, it was discovered the lack of fit the description of the coal swelling process with the use of one–parametric dependencies including those dependencies based on the theory of regular solutions on the solubility parameter of liquids. It was shown, that the quantitative connection between the swelling degree of coal and physical–mechanical properties of the solvents is achieved only on the basis of principle of the linearity of free energy under condition of taking into account the all solvation process. The basicity of the solvents and their molar volume are the factors determining the swelling degree for low–metamorphized coal.

References

[1] Kibler M. V. *Diejstvije rastvoritieliej na ugli* // In book: "*Khimiya tviordogo topliva*" Moscow, 1951. – v. 1. – p. p. 145 – 267.

[2] Keller D. V., Smith C. D. *Spontaneous fracture of coal* // "*Fuel*" – 1976. – v. 55. – № 4. – p. p. 272 – 280.

[3] Kröger K. *Die Steinnohleextraction* // "*Erdöl und Kohle*" – 1956. – Bd.9. – H.7. – s. s. 441 – 446.

[4] Sanada Y., Honda H. *Solvent extraction of coal* // Bull. "*Chem. Soe. Japan*" – 1962. – v. 35. – № 8. – p. p. 1358 – 1360.

[5] Sanada Y., Honda H. *Equilibrium swelling of coals in various solvents* // "*Fuel*" – 1966. – v. 45. – № 4. – p. p. 451 – 456.

[6] Sanada Y., Honda H. *Swelling equilibrium of coals by pyridine* // "*Fuel*" – 1966. – 'v. 45. – № 4. – p. p. 295 – 300.

[7] Kirov N. Y., O'Shea J. N., Sergeant G. D. *The determination of solubility parameters of coal* // "*Fuel*" – 1967. – v. 47. – p. p. 415 –424.

[8] Green T. K., Kovac J., Larsen J. W. *A rapid and convenient method for measuring the swelling of coals* // "*Fuel*" – 1984. – v. 63. – № 7. – p. p. 935 – 938.

[9] Mayo F. R., Zevely J. S., Pavelka L. A. *Extractions and reactions of coals below* 100 °*C* // "*Fuel*" – 1988. – v. 67. – № 5. – p. p. 595 – 599.

[10] Aida T., Fuku K., Fujii M. *et al. Steric requirements for the solvent swelling of Illinois № 6 coal* // "*Energy and Fuels*" – 1991. – v. 5. – № 6. – p. p. 74 – 83.

[11] Nelson J. F., Mahajant O. T., Walker P. L. *Measurement of swelling of coals in organic liquids* // "*Fuel*" – 1980. – v. 59. – № 12. – p. p. 831–837.

[12] Skrypchenko G. B., Khrennikova O. V., Rybakov S. I. // "*Khimiya tviordogo topliva*" – 1987. – v. 5. – p. p. 23 – 28.

[13] Osipov A. M., Bojko Z. V. // "*Khimiya tviordogo topliva*" – 1987. – v. 3. – p. p. 15 – 18.

[14] Van Krevelen W. *Chemical structure and properties of coal* // "*Fuel*" – 1965. – v. 44. – № 4. – p. p. 229 – 242.

[15] Larsen J. W., Shawyer S. *Solvent swelling studies* // "*Energy and Fuels*". – 1990. – v. 4. – № 1. – p. p. 72 – 74.

[16] Painter P. C., Graf J., Coleman M. H. *Coal solubility and swelling. Parts* 1, 2, 3. // "*Energy and Fuels*" – 1990. – v. 4. – № 4. – p. p. 379 – 397.

[17] Weyrich O. R., Larsen J. W. *Thermodynamics of hydrogen bonding in coal–derived liquids* // "*Fuel*" – 1983. – v. 62. – № 8. – p. p. 976 – 977.

[18] Marzec A., Kisielow W. *Mechanism of swelling and extraction and coal structure* // "*Fuel*" – 1983. – v. 62. – № 8. – p. p. 977 – 979.

[19] Szeliga J., Marzec A. *Swelling of coal in relations to solvent electron–donor* numbers // "*Fuel*" – 1983. – v. 63. – № 10. – p. p. 1229 – 1231.

[20] Hsieh S. T., Duda J. L. *Probing coal structure with organic vapor sorption* // "Fuel" – 1987. – v. 66. – № 2. – p. p. 170 – 178.

[21] Mayer U. Eine semiempirische *Gleichung zur Beschreibung des Lösungs–mitteleinflusses auf Statik und Kinetik chemischer Reaktionen. Th.* 1, 2. // "*Monutsh. Chemie*" – 1978. – v. 109. – H. 2. – s. s. 421 – 433; H. 4. – s. s. 775 – 790.

[22] Koppel I. A., Palm V. A. *The influence of the solvent on organic reactivity* // In: Advances in Linear Free Energy Relationships. Ed. N. B. Chapman a J. Shorter. London, New York: Plenum Press – 1972. – p. p. 203 – 281.

[23] Makitra R. G., Pyrig Ya. M. // "*Khimiya tviordogo topliva*" – 1988. – v. 6. – p. p. 41 – 45.

[24] Makitra R. G., Pyrig Ya. M. // "*Khimiya tviordogo topliva*" – 1992. – v. 6. – p. p. 11 – 20.

[25] Makitra R. G., Pyrig Ya. M., Vasiutyn Ya. M. // "*Khimiya tviordogo topliva*" – 1995. – v. 3. – p. p. 3 – 13.

[26] Aminabhavi T. M., Harogopadd S. B., Khinnavar R. S. *et al. Rubber solvent interactions* // "*Rev. Macromol. Chem. Phys.*" – 1991. – v. C 31. – № 4. – p. p. 433 – 497.

[27] Makitra R. G., Pyrig Ya. M., Zaglad'ko E. A. // "*Plasticheskije massy*" – 2001. – v. 3. – p. p. 23 – 27.

[28] Makitra R. G., Poliuzhyn I., Prystansky R., Smyrnova O., Rogovyk V., Zaglad'ko O. *Zastosuvannya pryncypu linijnosti vilnyh energij shchodo sorbciji ta pronyknennya*

organichnyh rechovyn // "*Praci naukovogo tovarystva im. Shevchenka*". – 2003. – v. 10. – p. p. 152 – 163.

[29] Makitra R. G., Prystansky R. // "*Khimiya tviordogo topliva*" – 2001. – v. 5. – p. 316.

[30] Larsen J. W., Green T. K., Kovac J. // "*J. Org. Chem.*" – 1985. – v. 50. – № 10. – p. p. 4729 – 4735.

[31] Hall P. G., Marsh H., Thomas K. M. *Solvent induced swelling of coals to study macromolecular structure* // "*Fuel*" – 1988. – v. 67. – № 6. – p. p. 863 – 866.

[32] Makitra R. G., Prystansky R. // "*Khimiya tviordogo topliva*" – 2003. – v. 4. – p. p. 24 – 36.

[33] Vasiutyn Ya. M., Makitra R. G., Pyrig Ya. M., Turovsky A. A. // "*Khimiya tviordogo topliva*" – 1994. – v. 4. – p. p. 66 –73.

[34] Makitra R. G., Pyrig Ya. M. // "*Khimiya tviordogo topliva*" – 1991.– v. 1. – p. p. 67–70.

[35] Makitra R. G., Pyrig Ya. M. // "*Khimiya tviordogo topliva*" – 1993. – v. 3. – p. p. 14 – 18.

[36] Makitra R. G., Bryk S. D., Palchykova O. Ya. *Doslidzhennya vzajemodiji malometarmophizovanogo vugillya z organichnymy rozchynnykamy (na prykladi diphenilolpropanu)* // "*Geologiya i geokhimiya goriuchyh kopalyn*". – 2003. – № 3–4. – p. p. 126 – 130.

In: New Topics in Monomer and Polymer Research
Editors: A. D'Amore and G. Zaikov, pp. 49-59
ISBN: 1-60021-436-3

Chapter 5

RECIPROCAL INFLUENCE OF SUCCINIMIDE AND OLIGODIENE- AND OLIGOOLEFIN-BASED SULFUR-CONTAINING ALKYLPHENOLS

R.Z. Biglova, A.U. Galimzyanova, R.F. Talipov, A.Ya. Gerchikov, V.R. Khairullina, G.E. Zaikov

Bashkir State University, Ufa; Emanuel's Institute of Bio- and Chemical Physics, Russian Academy of Sciences, Moscow, Russia

ABSTRACT

Reciprocal influence of antioxidizing and detergent-dispersive components of batch additives have been studied. Oligomers of piperylene and isobutylene functionalized by element sulfur and phenols in combination with succinimide have been chosen as the object of research. Constants of inhibition rates have been determined for oligomer antioxidizing agents.

Batch additives to various lubricating agents enjoy a steady increase in demand. The deterging components, succinimide for instance are used in oils with antioxidizing and anti-wear additives [1]. As separate additives possess operational properties to a different extent, the optimal overall quality is obtained at proper combination considering their effect and selection of appropriate compositions [2]. The effect of succinimide as a detergent, phenol-based oxidation inhibitors and chemical additives of polyfunctional action were studied for their washing and antioxidizing properties.

As antioxidizing reagents and additives of multi-purpose action phenol- and element sulfur-modified piperylene ($\overline{M}_n$=910) and isobutylene ($\overline{M}_n$=390; 880) oligomers: 4-oligopiperyleneylphenol, 4-oligoisobutenylphenol, 2,6–di-*tret*-butyl-4-oligopiperenylphenol, 2,6-di-*tret*-4-oligoisobutenylphenols; sulfided oligopiperylene (OPP), oligopiperelenyl-phenols and oligoisobutylenes (OIB); 2,2'-tiobis-(4-oligoisobutenylphenols) were used[3]. Succinimide was synthesized on oligoisobutylene ($\overline{M}_n$= 880) according to method [1], the content of nitrogen in it came to 3.98 mass %, that corresponds to *bis*-succinimide structure.

Washing action of imides of amber acid, whose molecule include groups of high polarity and hydrocarbon oleophilic part characterized by great affinity to oil was considered in the aspect of their solubilizing capacity, i.e. transfer by SAS micelle in a solution of products, insoluble in the given environment. Research into solubilizing ability of modified substances in isooctane environment was done by a technique described in [4].

According to [5], the concentration dependences of antiwear efficiency of sulfur-containing compounds have complex extreme character accounted for by micelle formation in oil solution of the additive. From this point of view it was interesting to investigate the concentration dependences of compositions in antiwear and detergent-dispersive additives.

In relation to the dye the solubilizing action was to a greater or lesser extent shown by all types of investigated compounds. The curves of the dependence of composition solubilization effect (SE) on additive concentrations are of a complex extreme character; various is also their arrangement in relation to the axes of coordinates. In experiments with rhodamine C mainly two types of curves (Fig. 1 a, b, c) were constructed: the first is characterized by fast colloid dissolution of dye and the solution stability (curves 2b, 3b); the other – by a longer stabilization period of solutions (curves 2a, 3a), but with a sharp growth of solubilization effect for compositions in the area of the higher concentration (curve 9 a).

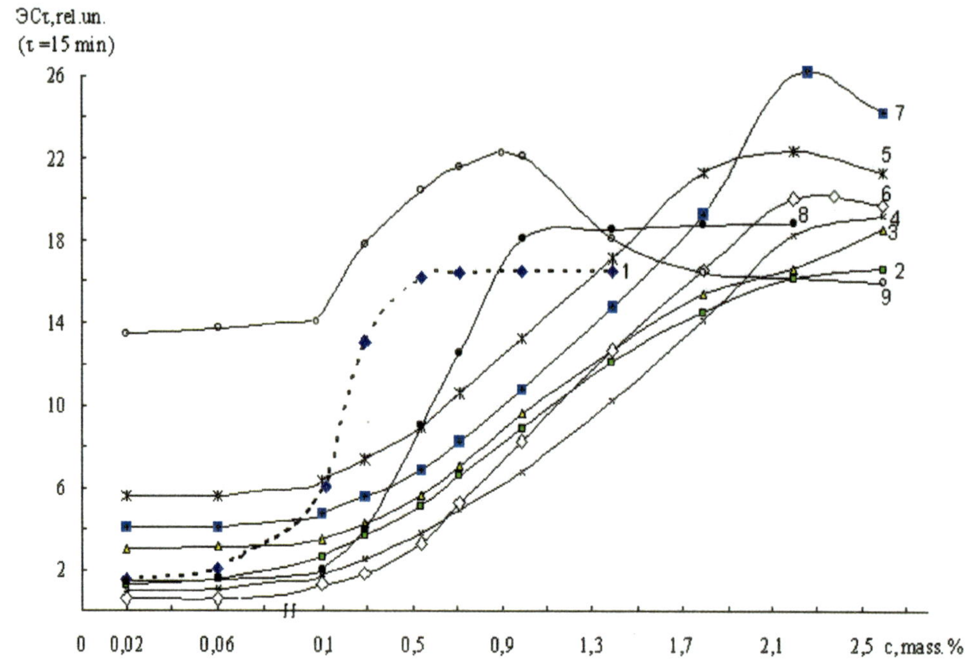

a) modified oligoisobutylene (M_n = 880)

Figure 1a. 1 – succinimide; 2 – sulfided oligoisobutylene; 3 – sulfided oligoisobylene-succinimide (0.3 mass %); 4 – 2,6-di-Bu^t-4-oligoisobutenylhenol; 5 – 2,6-di-Bu^t-4-oligoisobutenylphenol-succinimide (0.3 mass%); 9 – 2,6 – di-Bu^t-4oligoisobutylenphenol : sulfided oligoisobutylene, mass = 1:1; 7 – 2,6 – di-Bu^t-4-oligoisobutenylphenol : sulfided oligoisobutylene, mass. = 1 : 1 – succinimide (0.3 mass %); 8 – sulfur-containing *bis*-4,4`- oligoisobutenylphenol; 9 – sulfur-containing *bis* – 4,4`-oligoisobutenylphenol – succinimide (0.3 mass%).

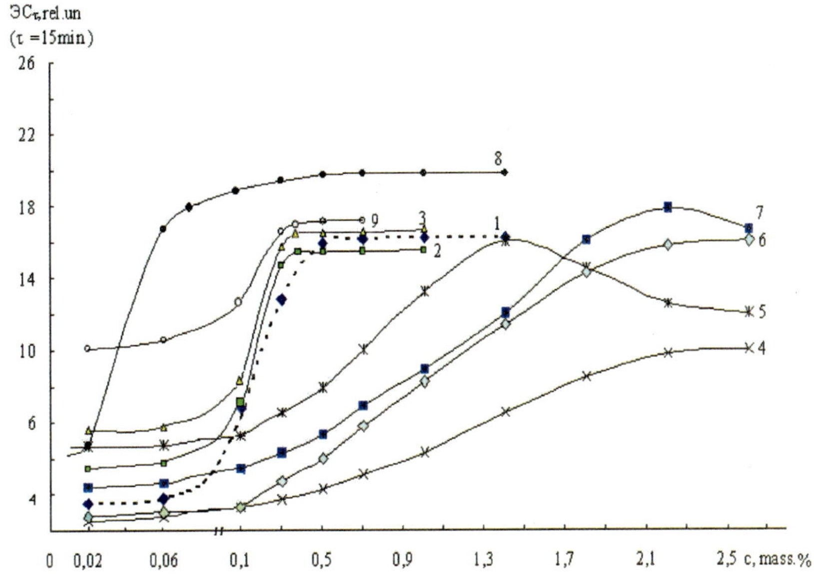

b) modified oligoisobutylene (M_n = 390)

Figure 1b. 1 – succinimide; 2 – sulfided oligoisobutylene; 3 – sulfided oligoisobutylene – succinimide (0.3 mass%); 4 – 2,6-d-Bu^t-4-oligoisobutenylphenol; 5 – 2,6-di-Bu^t-4-oligoisobutenylphenol – succinimide (0.3 mass%); 6 – 2,6–di-Bu^t-4-oligoisobutenylphenol : sulfided oligoisobutylene, mass. = 1 : 1; 7 – 2,6-di-Bu^t-4-oligoisobutenylphenol : sulfided oligoisobutylene, mass. = 1 : 1 - succinimide; 8 – sulfur-containing *bis*-4,4`-oligoisobutenylphenol; 9 – sulfur-containing bis-4,4`-oligoisobutenylphenol – succinimide (0.3 mass%).

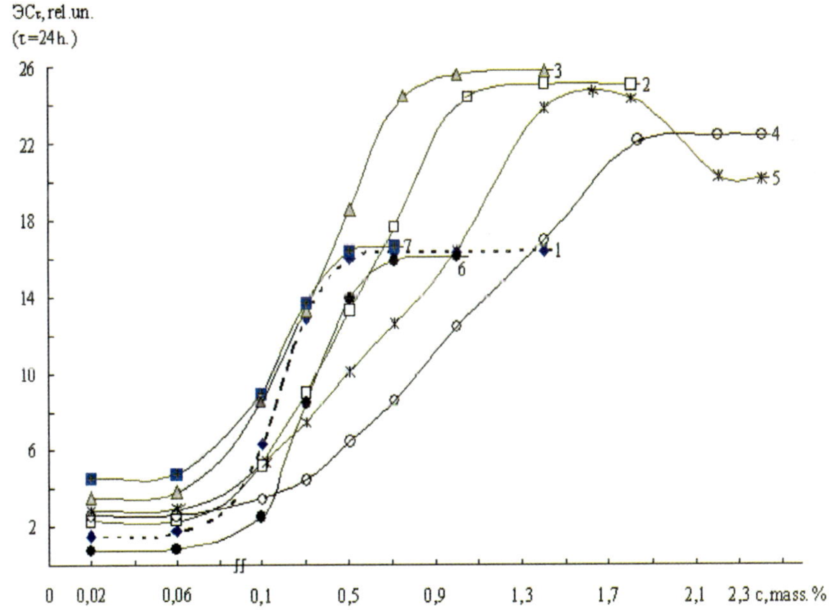

c) modified oligopiperylene

Figure 1c. 1 – succinimide; 2 – sulfided oligopiperylene; 3 – sulfided oligopiperylene – succinimide (0.3 mass%); 4 – 2.6-di-Bu^t-4-oligopiperylenylphenol; 5 – 2,6-di-Bu^t-4-oligopiperylenylphenol – succinimide (0.3 mass%); 6 – sulfided 2,2 –di-Bu^t-4-oligopiperylenylphenol; 7 – 2,6-di-Bu^t-4-oligopiperylenylphenol – succinimide (0.3 mass%);

Figure 1. The dependence of solubilizing ability on additive concentration in solution.

In compositions of these additives the solubilizing effect rises and critical values of concentrations for micelle formation (CCM) are shifted to the region of their reduction as compared to individual substances in isooctane (Table 1).

According to the Traube-Duclot rule influence on succinimide detergent-dispersive action of size and structure of hydrocarbonic radical of antioxidizing and antiwear additives introduced into oil composition is traced. The maximum solibilization effect takes place for the ramified and smaller radicals in size. Such effect is a property of substances synthesized on the basis of oligopiperylene ($\overline{M}_n$=910) and oligoisobutylene ($\overline{M}_n$=390). Experimental data (Table 1) testify that introduction of the synthesized substances in an oil composition with succinimide allows to substantially lower their dosage to maintain the operational characteristics at a sufficiently high level.

Table 1. Critical concentration of micelle formation in functionalized products, mass %

Additive	OPP ($\overline{M}_n$=910)	OIB ($\overline{M}_n$=880)	OIB ($\overline{M}_n$=390)
Succinimide	0.08 - 0.42	0.08 - 0.42	0.08 - 0.42
Sulfided oligomer	0.09 - 1.30	0.10 - 1.90	0.09 - 0.30
Sulfided oligomer - succinimide (0.3 mass. %)	0.08 - 0.70	0.08 - 1.70	0.08 - 0.20
Alkylated 2,6-di -Bu^t-phenol	0.20 - 1.80	0.30 - 3.00	0.20 - 2.20
Alkylated 2,6-di -Bu^t-phenol - succinimide (0.3 mass. %)	0.08 - 1.60	0.10 - 2.40	0.10 - 1.40
Product of successive alkylation and sulfiding in nonpolar solvents	0.09 - 0.56	0.10 - 2.40	0.16 - 2.20
Product of successive alkylation and sulfiding in nonpolar solvents - succinimide (0.3 mass. %)	0.09 - 0.30	0.10 - 2.20	0.08 - 2.10
Product of successive alkylation and sulfiding in polar solvents	-	0.15 - 0.95	0.07 - 1.00
Product of successive alkylation and sulfiding in polar solvents – succinimide (0.3 mass. %)	-	0.03 - 0.09	0.04 - 0.09

For further studies of succinimide effect on antioxidizing action of oligomers functionalized by sulfur and phenol, the latter were tested for the specified properties. Antioxidizing activity of the samples was measured on the rate of oxygen absorption in radical chain process of isopropylbenzene (IPB) oxidation with universal manometer apparatus (the relative error of reading is not more than 10 %) at the initiation rate of V_i = $7.21 \cdot 10^{-8}$ mol/L·s and temperature of 333 K in kinetic regime. Azodiisobutyronitrile (AIBN) was used to initiate the oxidation. The kinetic purity of IPB was estimated by comparing specially measured oxidation parameter $k_2/\sqrt{2k_6}$ with data given in literature (k_2 and k_6 are rate constants of oxidation chain growth and breakage considering cumylperoxile radicals of IPB). In practice this value is $k_2/(2k_6)^{1/2} = (3.2 \pm 0.3) \cdot 10^{-3}$ (L/mol·s) (the correlation factor r = 0.9953) [6]. The effectiveness of antioxidizing action of the samples was determined on the extent of the reduction of oxygen initial absorption rate in oxidizing the model substrate in sample presence.

To estimate reactivity of tested substances as AO the effective constant rate of oxidation chain breakage fk_{In} was used, where f – capacity of inhibitor action, k_{In} – constant rate of oxidation chain breakage on AO molecule. As a reference to be compared with a standard inhibitor – ionol, for which the value of fk_{In} = (2.0±0.2)·10^4 L/mol·s that is in good coordination with data [6], was chosen.

As the results of experimental studies show all modified OIB and OPP inhibit the process of initiated oxidation of IPB. The bigger starting concentration of samples entering the substrate under oxidation results in naturally lower oxidation rate. The dependences of starting rates of initiated oxidation of IPB on the concentration of the additives of substances under study are given in Fig. 2 and 3. In the radical chain oxidation mode of IPB the value of fk_{In} is employed in (1) [7]:

$$F = \frac{V_0}{V} - \frac{V}{V_0} = fk_{In}[InH]/\sqrt{2k_6V_i} \quad (1)$$

,

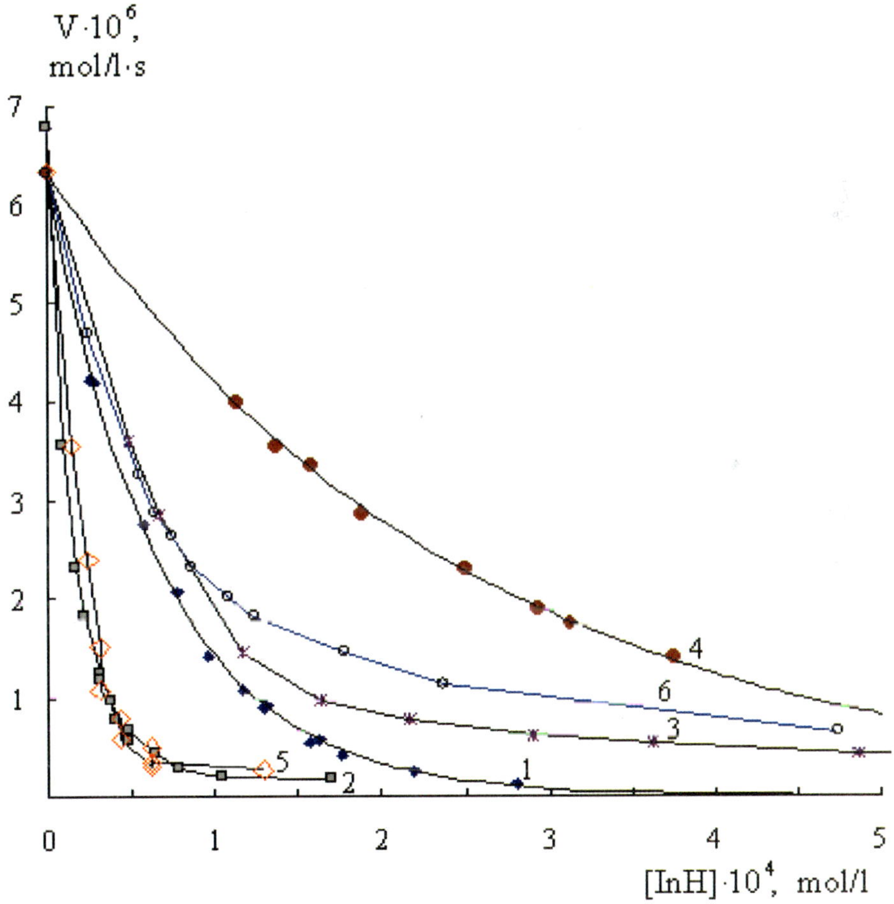

a)

Figure 2 a. 1. - 4-oligoisobutenylphenol ($\overline{M}_n$(OIB)=390); 2. - 2,2`-tiobis-(4-oligoisobutenylphenol), ($\overline{M}_n$(OIB) = 390); 3. - 4-oligopiperylenylphenol; 4. - Sulfided 4-oligopiperylenylphenol; 5. – Ionol.

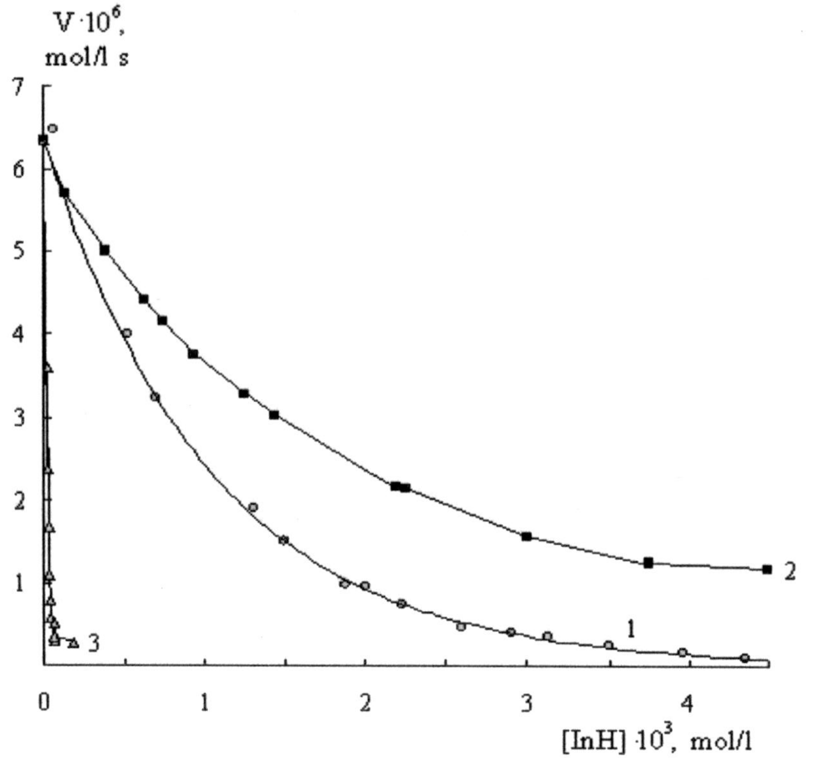

b)

Figure 2b. 1. - 4-oligoisobutenylphenol, ($\overline{M}_n$(OLB) = 880); 2. - 2,2`-tiobis-(4-oligoisobutenylphenol), ($\overline{M}_n$(OIB) = 880); 3. - Ionol.

Figure 2. The dependence of cumene oxidation rate on inhibitor concentration.

where v_0 and v are the starting rates of oxygen absorption in IPB oxidation in the absence and in the presence of AO respectively, V_i – the rate of oxidation initiation, [InH] – the concentration of inhibiting additive, k_{In} and $2k_6$ – rate constants of oxidation chain breakage on the inhibitor and quadratic chain breakage on cumylperoxile radicals of the substrate respectively [6]. Relationship (1) is established to satisfy all the studied substances (the correlation factor > 0.95, Fig. 2, 3). The linear dependence of inhibition parameter calculated in (1) on the concentration of AO under study testifies that in this interval of concentration quadratic breakage of inhibited IPB chain oxidation prevails.

Having processed the experimental data in the coordinates of equation (1) effective constants of inhibition rate fk_{In} were established. These values have been compared to the similar constants of chain breakage on the ionol molecule and are given in Table 2.

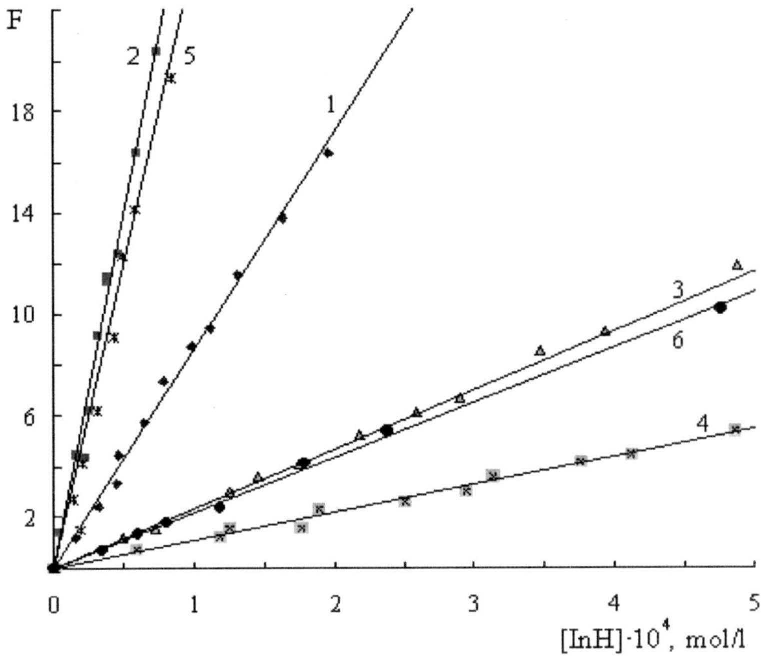

a)

1. - 4-oligoisobutenylphenol, ($\overline{M}_n$(OIB) = 390); 2. - 2,2`-tiobis-(4-oligoisobutenylphenol), ($\overline{M}_n$(OIB) = 390); 3. - 4 – oligopiperylenylphenol; 4. - Sulfided 4-oligopiperylenyphenol; 5. - Ionol; 6. – Succinimide.

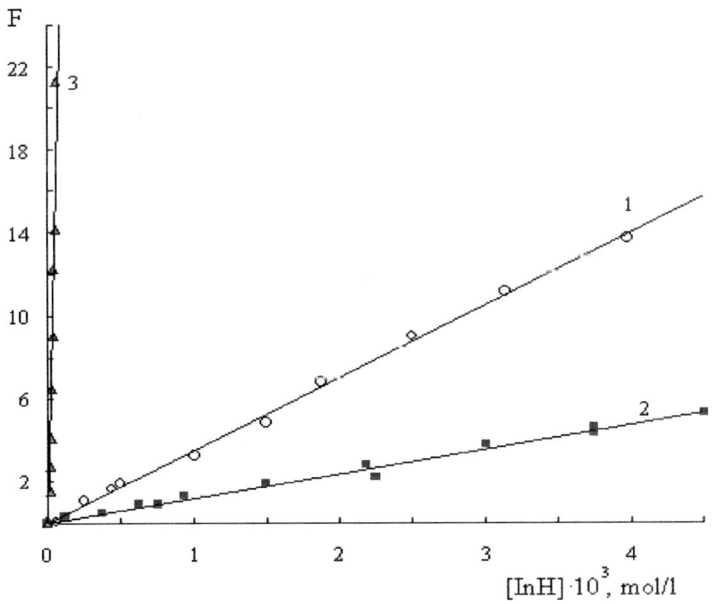

b)

1. - 4-oligoisobutenylphenol, ($\overline{M}_n$(OIB) = 880); 2. - 2,2`-tiobis-(4-oligoisobutenylphenol), ($\overline{M}_n$(OIB) = 880); 3. - Ionol.

Figure 3. The dependence of inhibitor parameter F on the concentration of inhibitor.

Table 2.

№	Inhibitor	$\overline{M}_n$(orig. oligomer)	fk_{In}·10-3, L/mol·s
I	4-oligoisobutenyphenol	390	5.30±0.50
II	2,2`-tiobis-(4-oligoisobutenyphenol	390	16.00±1.50
III	4-oligoisobutenylphenol	880	0.20±0.03
IV	2,2`-tiobis(4-oligoisobutenyphenol)	880	0.08±0.01
V	4-oligopiperylenylphenol	910	1.40±0.20
VI	Sulfided 4- oligopiperylenylphenol	910	0.70±0.10
VII	Succinimide	880	2.00±0.20
VIII	Ionol	220	20.0±2.0

This, antioxidizing activity of the tested compounds can be presented as a series:

IV < III < VI < V < VII < I < II < VIII.

As experimental data show succnimide also possesses a property to slow down the oxidation.

Interrelation between structure and antioxidizing activity of sulfur-containing phenol AO is of a complex nature, as inhibiting properties of these compounds are affected by the structure of phenol and sulfur fragments, their number and arrangement against each other. A certain correlation between the degree of intensity of inhibitor effect and the structure of AO molecule is observed: the length of alkyl chain, substitute character of sulfur atom, the number of sulfide and phenol fragments. In a series of oligomer alkylphenols the value of fk_{In} rises as the length of hydrocarbon radical (OIB, $\overline{M}_n$= 390) reduces and the number of phenol fragments grows by one macromolecule (OPP). Introduction of sulfur atoms in 4-oligoisobutenylphenol ($\overline{M}_n$(OIB) = 390) positively influence the oxidation parameters, that is being due to heterolytic decomposition of cumylperoxiles. In case of higher-molecular oligomers sulfur reduces inhibition rate constant, which is probably caused by steric factor and formation of stable radicals.

Combined action of mixtures of succinimide with piperylene and isobutylene oligomers modified by phenol and sulfur has been further investigated. It was shown that mixtures of AO III – VI with succinimide slow down oxidation somewhat more effectively than individual components in their composition. The dependences of cumene oxidation rate on the inhibitor content in the mixture with succinimide plotted in Fig.4 should be noted as the evidence of lack of any pronounced effect of succinimide on antioxidizing activity of ionol and OIB ($\overline{M}_n$=390).

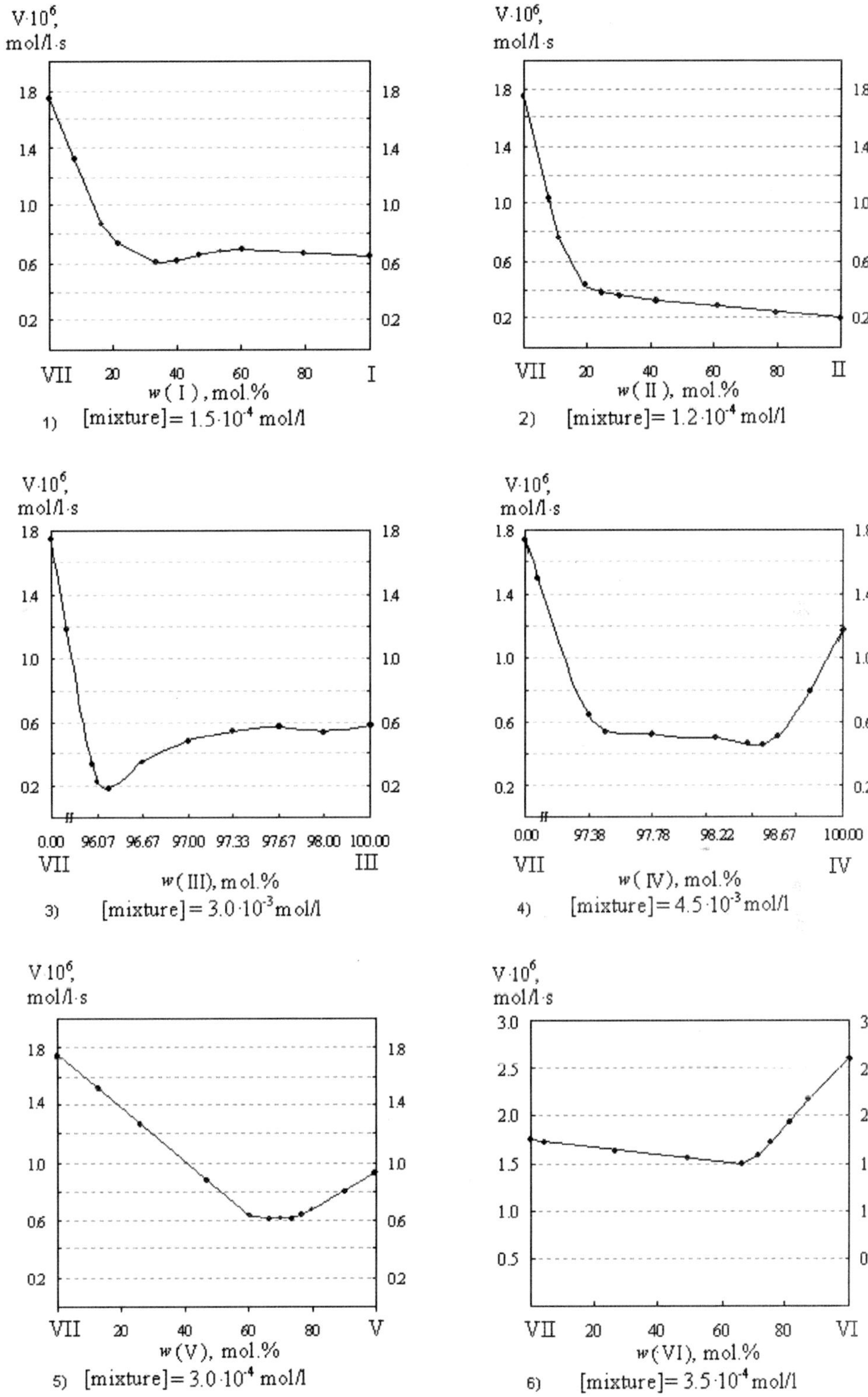

Figure 4. The dependence of cumene oxidation rate on the inhibitor content in the mixture with succinimide. Continued on next page.

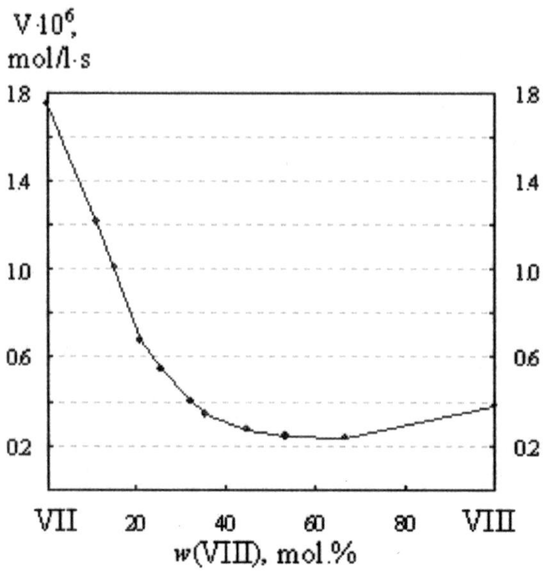

7) [mixture]= $1.2 \cdot 10^{-4}$ mol/l

Figure 4. (Continuation) The dependence of cumene oxidation rate on the inhibitor content in the mixture with succinimide.

Thus, the lack of antagonism in the examined binary mixtures allows to conduct the selection of optimal set of ingredients of batch additives on a smaller number of parameters.

Conclusions

1. Solubilizing ability of chemical additives on the basis of oligomers of piperylene (M_n = 910) and isobutylene (M_n=390; 880), their effect on detergent-dispersive action of succinimide have been studied; range of critical concentrations of micelle formation have been found. The presence of multifunctional additives allows to considerably decrease the dosage of washing agent in batch additives.
2. Antioxidizing efficiency of phenol sulfur-containing substances have been studied, fk_{In} for each one having been computed. Antioxidizing activity of sulfur-containing alkylphenols depends on the structure of sulfur-including fragment, the number and mutual arrangement of S-containing and phenol groups in their molecules.
3. Succinimide in oily composition exerts no marked effect on antioxidizing activity of isobutylene oligomers (M_n=390) modified by element sulfur and phenol.

REFERENCES

[1] Glavati O.L., Popovich T.D., Borislavsky B.T., Zhurba A.S., Lukashevich A.N. The improvement of quality of cuczinimid additives// *Khimiya i tekhnol. topl. i mas*.- 1989. – № 3. –S.22.

[2] Selezneva I.E., Levin A.Ya., Monin S.V. Detergent-dispersion additives to motor oils // *KhTTM*. – 1999. – № 6. – S.39.

[3] Biglova R.Z., Malinskaya V.P., Minsker K.S. Polimer-analogous reactions of polyolefins with phenols and aminophenols // *Vyisokomol. soed*. – 1994. – T. 36 A. – № 8. – S.1276-1280.

[4] Bipper A.B., Krein S.E., Sher V.V., Sanin P.I. Solubilizing action of additives of various structure and its effect on motor oil properties // *Neftekhimiya*. – 1968. – T.5. – № 5. – S.798.

[5] Loznetsova N.N., Pavlov K.A., Toporov Yu.P., Shchegolev G.G. The role of micelle formation in anti-wear properties of lubricating oils // Proceed. of 2nd Intern. confer. *«Colloid-2003»*. 2003.- S.184.

[6] Denisov E.T. Rate constants of hemolytic liquid-phase reactions. M.:*Nauka,* 1971. – 700 s.

[7] Denisov E.T., Azatyan V.V. Inhibition of chain reactions. *Chernogolovka*. – 1997. – 136 s.

In: New Topics in Monomer and Polymer Research
Editors: A. D'Amore and G. Zaikov, pp. 61-73
ISBN: 1-60021-436-3

Chapter 6

MODERN CONCEPTION OF BIODEGRADABLE AND BIOCIDE POLYMERS AND COMPOSITE MATERIALS ON THEIR BASIS

Olga A. Legonkova
Moscow State University of Applied Biotechnology, Russia
e-mail: OALegonkovaPB@mail.ru

There are several definitions used for description of biological destruction. They have no significant difference [1,2,3], all saying that biodegradation is a process caused by biological activity that leads to generation of natural metabolic end-products with the change in chemical structure of material.

Biodegradable polymer materials can be subdivided into three groups.

1) plastics on the basis of natural biopolymers (natural rubber, proteins, polysaccharides, chitin, epoxidized oils, unsaturated vegetable oils polymers, lignin, polluan etc.) [4].
2) chemically and biologically synthesized polymers and their mixtures. The major stress in the sphere of biodegradable polymers and co-polymers is made on the variation of monomer structure. Great opportunities associated with the changes in synthesized copolymers' microstructure are merely unrealized. Polyhydroxialcanoates - *Flcaligenes eutrophus, Flcaligenes eutrophys, Alcaligenes eutrophus* bacteria [5] (in fact polyhydroxyfatty acids are produced by more than 100 species of microorganisms) are synthesized microbiologiacally as well as polylactid PLA – microbial fermentation of cereal processing waste products. Pure cellulose can be obtained from low molecular glycosides under the influence of microorganisms of Acetobacter familia Sarcina ventriculi, Shaerotilus natans, Lampropedia hyaline [6,7]. With the use biotechnical processes 1,3-propandiol, epichlorhydrine are obtained. However biotechnological processes of high-molecular substances are not well described in scientific literature.

Among synthetic polymers the major biodecomposed ones are aliphatic-aromatic polyesters, aliphatic poliesteruretanes (having more than 75% polyester segment on the basis of aliphatic acids: siccine, adipine, threemellithic) and glycols: ethylene glycol, butanediol, dimethylciclohexane), polyamides (containing more than 30% of ether fraction, amide fraction is obtained in polymer in the form of polycaprolactame and laurinelactam blocks), polyester carbonate (containing more than 70% of ether fraction) their copolymers, polyvinyl alcohol, a number of polymers, Table. 1 [2,7-12]. The ways of biodecomposed polymers' synthesis are published in [13].

Generally polymers with irregular group distribution are better decomposed: acetalic compounds (starch, cellulose), compounds with amid groups (peptides, proteins), ether groups (polihydroxifatty acids). Low MM (Molecular Mass), low crystal rate and high hydrophilic qualities are the factors enhancing decomposition ability.

Table 1. Microorganisms and enzymes decomposing polymer materials

Polymers	Average molecular mass	Microorganisms and enzymes decomposing polymer materials
Polyvinyl alcohol	20000-90000	bacteria *Pseudomonas SP*
Oligopolystirol	400	*Alcaligeues S.P.*
Oligopolybutadien	650	*Acinetobacter S.P.*
polyacrylonitrile	160	*Fusariumbacter, Arthrobacter*
Polyethylene glycol	400-2000	Soil bacteria, *Freudomonus*
polypropylene glycol	4000	Soil bacteria
Poly-β-methylpirilacton	3000	*Allenespadonxua*
Adipates (ethers of adipi(ni)c acid)	300-850	lipases, Soil bacteria
Oligomer of plyamide-6	Rate of polymerisation 1-6	*Corynebacterium aurantiacum, achromobacteruttatus*
Poly-L- glutamic acid	4000-100000	*Scopalariopsis brevicaulis, protease*
Polyprolacton	25000	Soil bacteria
Olygomers of polyacrylics	500-1000	*Xanthamanasmalto-Phila, Acinetobacter*
Polysrirol olygomer	400	*Alcaligenes sp.*
polyethylenes	5000	bacteria
polyurethanes	1000-8000	Different moulds, *Apergillus niger, Chactomium globusum*
polyamide -6	16000	*Corynebactecium aurdntacum, Achromobacter duttatus,Flavobacterum sp.*
Poly-L- lysine	75000-200000	*Scopulariopsis brevicaulis, protease,*
Composite of polyethylene- polyvinyl alcohol copolymer with starch		*Penicillium funiculogum*
Composite of polypropylene-maleic anhydride copolymer with starch		*Penicillium funiculogum*
Polyamide		*Cladosporium herbarum, Acrodontium crateriforme*
Natural rubber		*Actinomycete Nocardia, Penicillium*
lignine, cellulose		*Penicillium, Trichoderma, Mycor, Aspergillus, Chaetomium globosum kunze, Coniophore cerebella, Coriolus versicolor*
Epoxy resins, polychlorvinyl plastic		*Stachybetrye lobulate, Syncephalastrum racemosum, Papularia sphaerosperms, Monosillium exsolum.*

3) composition materials. At first biodegradable polymers were mixtures of common well-known polymers (ethylene-vinylacetate copolymer, ethylene-propylene copolymer, ethylene-vinyl alcohol copolymer, ethylene-acryl acid copolymer and linear polyurethanes) with starch. Polyethylene starts its biological decomposition in the presence of starch [14]. There is a process of biodegradation, resulting from four

processes: water diffusion, reduction of specific surface area because of pore-formation, microorganisms' adsorption on the surface, fermentative hydrolyses of the material [15-21].

However the time has shown that such mixtures are not fully biodecomposed: starch breaks up under the influence of microorganisms and the polymer matrix itself is fragmented.

The ratio of the components in the composite material can influence the rate of biodegradation. Undoubtedly biodegradable polylactide accompanied by polyvinylacetate (more than 30%) doesn't practically undergo biodegradation [22].

Now there is a range of new biodecomposed composite materials gained applying different technologies by 20 leading companies of the world [5,23,24].

Modern research work in the field of biodegradable materials is carried out in two directions:

- development of polymer materials and auxiliary substances that are quickly decomposed and fully mineralized under the influence of microorganisms. The polymers can be obtained both from petrochemical and renewable stuff e.g. production from polyvinyl alcohol and natural starch.
- development of polymer materials and auxiliary substances that should not be quickly decomposed. Classical examples are rubber, casein polymers and cellulose acetate. Cross-link chains are also a new product on the basis of polysaccharides and proteins for isolating materials and woodchip boards.

Recently polymers and auxiliary substances from renewable resources become more and more popular in different areas of usage. The advantages of saving petrochemical resources and cut in the amount of waste products are closely connected with the creation of various biodegradable materials of better and better quality [8].

Consequently, three key elements are necessary for full biodegradation of polymer material: presence of microorganisms, selectively influencing the polymer materials, polymer materials themselves and necessary environment, table 2. If one of the elements is missing biodegradation doesn't take place. For example newspapers and food stuff that remain in their initial state after staying in the soil or spoil areas.

Table 2.

Biodecomposition		
Microorganisms	Material	Environment
-corresponding enzymes -necessary concentration of enzymes, -co-metabolism, -aerobic, anaerobic -inhibitors/initiators -outer- и intro-cell influence mechanism, - organisms inhibiting enzymes' action	-chemical compounds - presence of branching -hydrophilic/hydrophobic - stereochemistry, -molecular mass, -chains' mobility, -degree of crystallinity -interaction with copolymer -surface -alternative carbon sources	-temperature, -oxygen, -humidity, -salts, -metals, -residual nutrient media - pH rates, -pressure, -light

A polymer, exposed to decomposition should, satisfy certain necessary requirements. [9]:

- a polymer chain should contain objects exposed to hydrolysis and oxidation. The polymer containing not more than 10 carbon atoms in monomer chain is considered the most resistant (stable).

Available links: **R = CH$_2$; R = CH – R' ; R – CH –OH ;**
R – CHOH – R ; R –CO - H ; R – CO – R' ; R – CO - R

A significant reduction of ether stability is observed when there are more than 12 carbon atoms in a chain:

- the rate and type of branches play an important role because, for example, new special enzymes may be necessary for branching substances. The ratio between hydrophobicity and hydrophillity causes extra influence. There could be positive effects at the same time e.g. the hydrolysis speed correlates with hydrophillity of the material.
- many enzymes have selective influence upon stereoisomers. This particularity can be used for depolymerisation kinetics control.
- spreading of molecular mass has a key influence upon the speed of depolymerisation. It is also influenced by enzymes' availability, chains' mobility, rate of crystallinity (the more the rate of crystallinity the less the ability to hydrolysis as crystalline areas are less accessible to enzymes) and other morphologic properties.
- interaction with other polymers should be taken into account, as they can act as a barrier layer during the moisture and gas diffusion.
- it is possible that there is influence of other chemical (hydrolysis, oxidation) and physical processes (leaching, sunlight, photodestruction, weather, mechanical abrasion) upon biodecomposition.

Various factors described above when combining their actions make it more difficult to reconstruct the conditions for experiments or in technical devices.

Resistance of polymer materials to microorganisms' activity also depends upon plasticizers used, fillings, stabilizers, and technological additives, their ability to be a source of carbon, nitrogen and other biogenic elements for microorganisms. It is well known, that inorganic components – silicates, sulphates, phosphates, carbonates don't enhance fungi growth. It is demonstrated that fungi-resistance of polyvinylchloride samples doesn't indicate their bacteria resistance. For example slate polyvinylchloride is fungi-resistant, but is well-used by denitrifying and carbon-containing bacteria.

Nowadays, there are no polymer materials fully resistant to the activity of living beings. The research work carried out in Japan showed that such materials as neoprene and nylon. They are not affected by moulds according to American data, become worse under the influence of *Aspergillus niger, A. Flavus.* There are different opinions among scientists considering other polymers: kapron, poluuretane, cellulose acetate, cellulose acetobuterate, phenol-aniline-phormaldehyde resin, melamine-phormaldehyde resin [25].

Environmental factors should be selected in such a way that microorganisms had optimal conditions for biodecomposition e.g. temperature, humidity in liquid and gas phase, salts (type and concentration), oxygen presence or absence (aerobic or anaerobic decomposition), availability of alternative electron acceptors, microelements, nutrients, pH rates, redoxpotentials, stability or changes of environment, pressure, microorganisms "rivals" (protozoa), inhibitors, alternative carbon sources, light (intensity, wave length, cycle). At the same time a minimum water content is necessary [10,26-29].

In all the cases of biodamage there is interaction between living organisms and polymers and polymer composite materials [1]. Consequently, the problem of biodamage is ecotechnological.

Biodamage is caused by:

1) microorganisms (animals, higher plants)
2) microorganisms (bacteria, fungi). The most aggressive are moulds of genus Aspergillus, Penicillium, Trichoderma. According to the type of polymer, microorganisms and environmental conditions the mechanism of biodecomposition can vary. Mould spawn can use tiny cracks and pores of the material in the juncture between plastic and components' particles for its growth.

There are three type of microorganisms' influence upon polymer materials:

1) mechanical,
2) metabolic product's influence (organic acids, enzymes, amino acids, pigments) upon chief physico-chemical and technological properties of the materials.
3) biopollution of polymer materials and things made of them.

Mechanical destruction of polymers takes place due to fungi hyphae mycelium growth (the same as grass growing through asphalt). Mould spawn can use very thin cracks and pores of the material for its growth.

Biopollution appears because of spores, conidium or single parts of mycelium presence in different goods. In the process of their vital activity moulds and bacteria can secrete a large number of metabolites, that have negative influence on the materials' properties. Oxidative fermentation caused by mould fungi and oxidative bacteria is possible because of the peculiar oxidation-reduction enzymes present in microorganisms.

The influence of metabolic products enhances two major processes leading to biodegradation – hydrolysis and oxidation.

Thus metabolism of polymer decomposition products on the basis of lactic acid is well-researched. On the initial step and out of living organisms biodecomposition takes place by pure hydrolysis with the lowering of molar mass. Decomposition in acid area (pH<1,5) is negligible. In the alkaline area (pH>7,5) hydrolysis is fast. In the human body decomposition runs faster, than while composting. The following □Øzymes are responsible for biodecomposition: propase, proteinase K, bromehaine, phicine, esterase and trypsyne.

The reactions of microbiological transformation of carbohydrates are in general oxidative processes, Fig. 1-4. Alcohols, aldehydes, keto- and oxyacids are produced and then oxidized and fermented [7, 17-34].

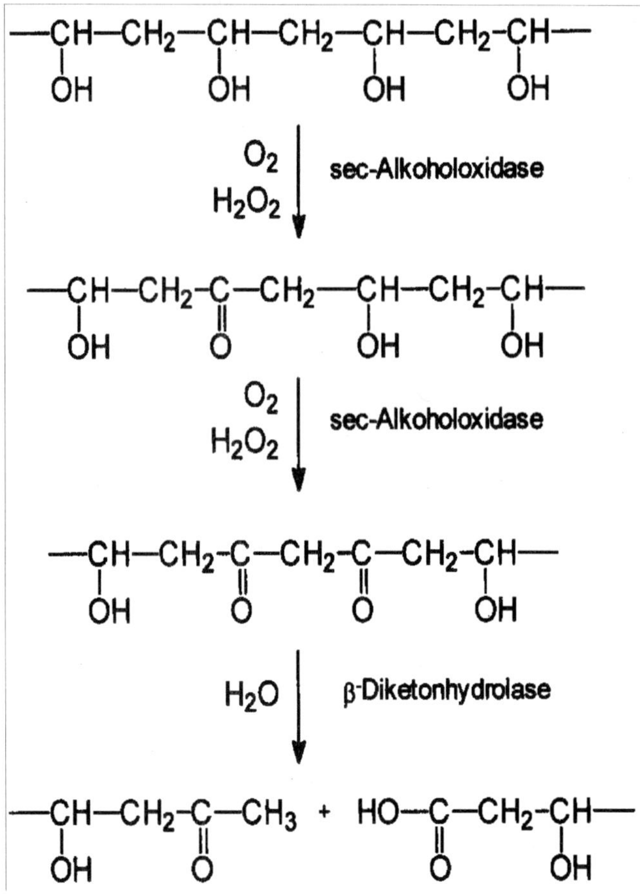

Figure 1. Mechanism of fermentative depolymerisation of polyvinyl alcohol.

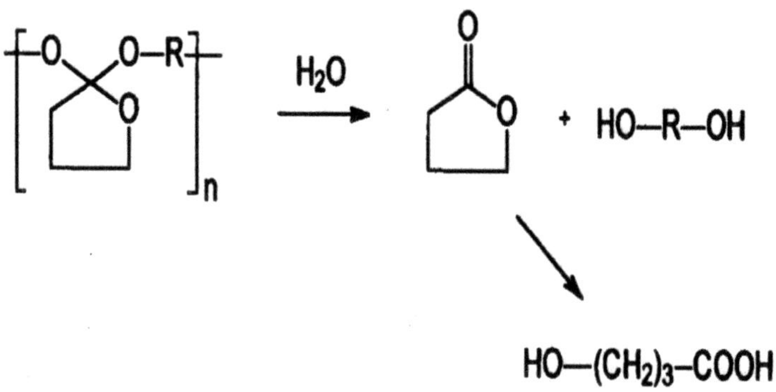

Figure 2. Mechanism of fermentative decomposition of polyortoether.

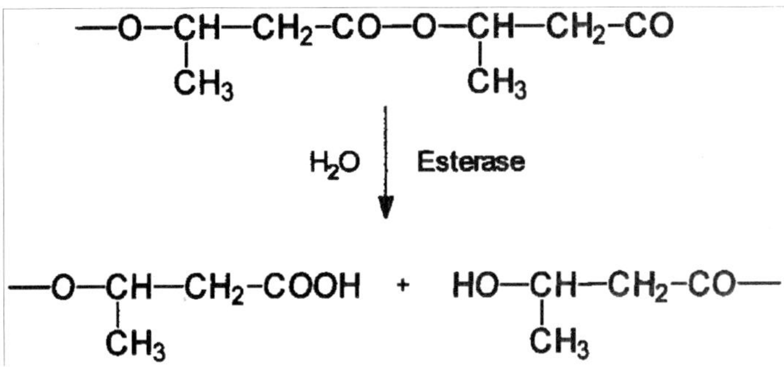

Figure 3. Hydrolytic and fermentative decomposition of polyhydroxifatty acids.

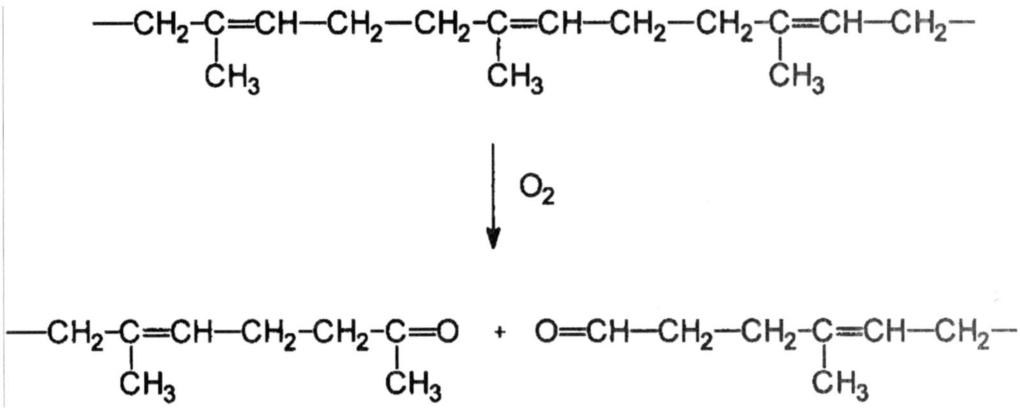

Figure 4. Enzyme depolymerisation of natural rubber.

During the oxidation of β-carbon atom of alkan (hydrocarbon) chain alcohols and ketones are produced. Ketone degradation results in primary alcohol having a chain that has 2 carbon atoms less than the initial substrate. This alcohol can then be subjected to oxidative dehydrogenation with the following β-oxidation of generated fatty acid [35,36].

Oxidation of non-final hydrocarbons under the influence of enzymes, secreted by microorganisms, takes place with alcohol, aldehyde, keto- and oxyacids generation many of which are then transformed into dibasic then subjected to β-oxidation [35, 36].

The splitting of aromatic hydrocarbons is followed by the production of fenols that then oxidize to neutral dibasic. Alkilreplaced aromatic compounds are subjected to benzyl oxidation by fungi Aspergillus niger, Aspergillus sclerotiorum, Penicillium adametri [37].

The splitting of cellulose by microorganisms lead towards the creation of olygo- and monosaccharides, CO_2 [38], polyimide splitting leads to the destruction of imide's cycle. In the case of mylar changing of the consumption stripes in the areas 1720 cm^{-1}, 1330 cm^{-1}, 990 cm^{-1}, 730 cm^{-1} (corresponding to the groups C=O, CH_2, C-O, benzene ring, respectively) [34].

It is stated that acetic, propionic, butyric, fumaric, siccine, malic, citric, tartaric, gluconic and oxalic acids are present in the fungi metabolites. Organic acids have a double role: on the one hand they act as aggressive media to plastics that can lead to changes in physico-chemical characteristics of the materials; on the other – they are a source of carbon for fungi further development [39-50].

Plastics are affected by the following microorganisms: fungi - *Asp. niger, Asp. versicolor, Asp. flavus, Asp. amstelodamii, Asp. ruber, Pen. purpurogenum, Pen. brevi-compactum, Pen.*

commune, Pen. palitans, Cladosporium, Fusarium, Paccllomyces Penicillium sp., Aspergillus wamori, Aspergillus oryzae, Trichoderma sp., Chaetomim globosum, Trichoderma lignorum, Cephalosporum aeremonrum, Rhizopus nigricans, Alternaria, bacteria - *Ptreptococcus rubrireticuli, Pseudomonas aeruginosa, Streptococcus rubrireticull* [51-60]. *Fspergillus niger, Candida, Lipolitica, Sporatrichum sp.* fungi can realize the olefine oxidation by reactions of three types: 1. reaction of hydratation of double bond with the generation of monobasic alcohols; 2. reactions that result in creation of epoxies; 3. reactions summing in the generation of diols.

Fungicides (generally with the melting point above 160°C) are used for the preservation of plastics: salicileanylide, zinc-salicileanylide, copper 8-oxyhinolade, phenol derivatives, quaternary ammonium compounds, trylane, trychlorethylphosphate, 1-flourine-3, bromine-4,6, dinitrobenzene, tryethylmetacriloxistanane, mercury salicilatephenyl, mercury ftalatephenyl, pentachlorphenyl, salicyl-anylidee, copper 8-oxinolate etc. [4, 52, 61-63]. The following compounds are used as antiseptics for lumber: sodium fluoride, dinitrophenol, wood resin, ammonium tetraphtorborate, bolidene, copper naphthenate, donalyte [64]. The authors [65-68] demonstrate that iron dioxide in the structure of composite polymer materials stimulates the growth of microorganisms, titanium dioxide is inert, and zinc oxide inhibits it. Among the fillings asbestos and talc enhance and calcium carbonate inhibits the growth. Low fungi resistance of a number of composites is connected with the presence of magnesium oxide as a filling in their composition that absorbs moisture, swell and thus promotes the development of microorganisms. Metalloorganic, chloroorganic quaternary ammonium and inorganic compounds classes of biocides were used to protect polymer materials from biodamage [49].

Fungicides with a limited time of toxic action are of special interest because their toxicity is not a problem any more when their time of action is up. These fungicides can be subdivided into the following groups: 1. inorganic compounds (ZnO, $CuSO_4$, NaF, NH_4BF_4, $Ca(CN)_2$); 2. organic compounds (phenol derivatives, phosphorous-containing compounds, amines etc.); 3. metallorganic compounds containing Hg, Pb, Sn, Na, Ag; 4. multifunctional additives, that are in fact synthetic biologically active compounds (regulations of the biodecomposition time is based upon the phenomena of toxotropic reduction of inner tensions because of realization of mezomorphic permolecular level of organization, promoting a large number of functional groups taking part in the adgezine and cogezine interaction, and their even local distribution in the system [69]).

It is known that microorganisms are present everywhere: soil, water, air, plants, animal and human bodies [70]. In all cases of biodamages there is interaction between living organisms and polymers. The problem of biodamages is eco-technological.

However soil is the major habitat of microorganisms [71,72]. Their genus variety and their quantity depend on the type of soil, its structure, fertility and other reasons. The richest layer is 5-15 cm deep, 1g. of the soil here contains up to 10^8 microorganisms. Usually the number of microorganisms directly depends on the amount of organic waste products in the soil [73].

The soil metabolism depends on the interaction in the bacteria community. There is a striking variety of microorganisms in every type soil, sometimes they have opposite and incompatible to the same habitat qualities [74].

The organic substances decomposition process in the soil is performed by successive reactions with the participation of different microorganisms, changing each other and delivered from the soil's great storage – microbial pool [75, 76]. Cosmopolitan microorganisms of all soil types are fungi of genus Penicillium, Aspergillus and bacteria Bacillus mycoides, Bacillus megaterium. They have one common quality – the ability to secrete enzymes. No substance is created without enzymes in nature. Every splitting of a dead organic substrate is a fermentative process resulting in simple chemical compounds' release, that are then utilized both by microorganisms and superior plants.

The successful solving of the problem is mostly connected with the knowledge of microbial damage real process and the presence of objective quantitative information of its rules and processes.

In spite of the fact, that nowadays there is a large number of products in the market of biodecomposed materials, the opinions at the biodecomposed polymers are contradictory [78]. There is a marking for biodecomposed materials, fig. and approved examination standards.

DIN 54-900 – "Checking polymers' composting ability". The test shows if a polymer material could be biologically decomposed in the conditions of controlled composting. The standard is based upon multi step test scheme and is divided into 3 parts:

DIN 54-900-1 – chemical testing of polymers. We determine:

- elementary check for C, H, O, S, N;
- content of organic dry particles, total carbon content and content of organically bound carbon;
- general nutrients;
- inorganic harmful substances (polychloric byphenyls, polychloric dibenzodioxines or dibenzophuranes).

Materials are not considered valid if the mass part of all organic components is less than 50%. Content of harmful substances should be so low that it wouldn't hinder the recycling in the compost.

DIN 54900-2 – check of polymers complete biological dissolubility in the laboratory experiment. Three methods can be used here.

- determination of biological dissolubility in the water media by determining the biological need for oxygen in respirometer.
- determination of biological dissolubility in the water media by determining the CO_2 excretion and creation of carbon balance.
- determination of biological dissolubility in compost by determining the CO_2 excretion and remaining polymer in compost.

We can speak about complete decomposition if minimum 60% of organic carbon is recycled during maximum 6 months.

DIN 54900-3 – practical checks. The material under research should be tested according to the stated methods under the conditions close to practice. Test is done in biocompost according to methods allowing airing and shoveling. The samples are mixed with fresh biological waste products and placed into bags. The content of the bags is analyzed regularly, first bolted throw 10 mm sieve, then from 2 to 5 mm and at last less than 2 mm. The remaining parts of biodecomposed material are sorted and checked. The sorted particles are dried, weighed and the organic amount is figured out by piercing. Moreover the maximum thickness of the decomposing layer can be determined.

ASTM G 21-90: a method for synthetic polymers' fungi-resistance determination.

DIN 40 046: climatic and mechanical testing of electric components and communication devices. The sample is hanged in the closed glass vessel with water on the bottom. The mould formation is determined visually. The mass reduction can be determined and also durability to stretching.

DIN 53 739 (1993): polymer check under the influence of fungi and bacteria, visual check of mass and physical properties changes.

ISO 846 (1987): determination of behavior under the influence of fungi and bacteria – checks by means of visual control and measurement of mass and physical properties changes.

Composting is according to DIN 53 739 a test performed by digging into the soil. The samples are dug into the mixture of garden compost and greenhouse soil (humidity 60%) – maximum ability to hold water (pH 7,5). The rate of the soil germination is evaluated by the total number of bacteria - $*10^8$ KOE/g. fungi - $*10^4$ KOE/g. at the temperature 28°C, air humidity 90-100% for 28 days. Soil activity control is made on cotton linen; its toughness should rise more than 75% for 7 days. The samples have the form of plates 30x30x0,8 mm in size. To determine the mass changes of the tested materials the samples are dried by silica gel for mass permanency and weighed with the accuracy of 0,1 mg. After the incubation the samples are cleared, put into the exicator till the mass is constant, then weighed with the accuracy of 0,1 mg.

Storm-test, ASTM 5209-91 is widely used to estimate biodecomposition.

GOST 9.049-91is in action in Russia "Polymer materials and their compounds. Laboratory tests methods for mould fungi resistance".

There is a list a requirements that are applied to biodecomposed materials:

1) biodecomposed materials should not have negative influence upon technical and technological processes, gas exchange, thermophilic phase and/or hygienisation phase and other;
2) biodecomposed materials should decompose completely. After this pieces of biodecomposed materials should not be visually distinguished and parts of biodecomposed materials should not be found after second extraction.
3) biodecomposed materials should not influence the compost quality and its recycling.

However, in spite of all this, nowadays data about biodecomposed polymers and materials on their basis are very fragmentary and are merely the results of standard research works of the materials' microbial resistance.

REFERENCES

[1] A.A.Anisimov, V.F.Smirnov. *Biodamage in industry and defence from it.* 1980, 80 p.

[2] O.A.Legonkova, O.A.Sdobnikova, A.L.Peshehonova, A.A.Bokarev. Biodagradable materials in packaging technology. *«Tara i upackovka»,* 2003, № 6.

[3] L.A.Sukhareva. *Biostabel coatings.* Pensa, 2000, 45 p.

[4] O.A.Legonkova, L.A.Sukhareva. Thousand and one polymers from biostable to biodagradable, UPGroup, Moscow, 2004, 270 p.

[5] O.A.Legonkova, O.A.Sdobnikova, A.L.Peshehonova, A.A.Bokarev. Biodegradable materials. *Biodegradable materials. «Tara i upackovka»,* 2004, №6.

[6] Fonken G., Gonson V. Microbiological oxidation, Moscow, *Mir,* 1976

[7] Kaplan D.L., Mayer J.M., Ball J.D., McCassie J., Allen A.L., Stenhouse P. Fundamentals of Biodegradable Polymers. In: Ching Ch., Kaplan D.L., Thomas N.: Biodegralable Polymers and Packaging. *Lancaster,* 1993, pp.1-43.

[8] Wolfram Tanzer. Biologisch abbaubare Polymere. Deutscher Verlag fur Grunstoffindustrie. Stuttgart, 2000, 200s.

[9] Aminabhavi t.V., Balundgi R.H. A review on biodegradable plastics, *Polym.-Plast.Tchnol.Eng.,* 1990, 29 (3), pp. 235-262.

[10] Shagal D.D., Agarwal P.N. Deterioration of paints by microorganisms. Zabdev. g. *Sci. andTechinol,* 1968, 6, cc.205-211.

[11] Huang S.I. The effect of structural variation on the biodegradality of synthetic polymers. *Amer. Chem Bac. Polym. Prepr,* 1977, 1, pp. 438-441.

[12] Alberson A.C., Ranby B. Biodegradation of Synthetic Polymers: The C-14 Method Applied to Polyethylentes. Proceedings of the Third International Biodegradable Symposium. London: *Applied Science,* 750 p.

[13] Vasnev I.I. Biodagradable polymers. *VMS,* 1997, Serjia B, v.39, №12.

[14] Grfiin G.J.L., Mivetchi H. Proc.3d. Int.Biodeg.Symposium. Kingston, 1975, 807 p.

[15] Kasjmina N.A. Development of composed thermoplastic materials based on starch raw materials. Moscow, 2002.

[16] Patent 2174132, Russia.

[17] Patent 2126427, Russia

[18] Patent 2126023, Russia.

[19] Patent 2123014 Russia.

[20] Patent 2117016 Russia.

[21] Sychugova O.V. Structure and biodegradation of ethylene-vinylacetate composition with thermoplastic starch. Moscow, 2004.

[22] Formin V.A., Guseev V.V. Biodegradable polymers, state and perspectives of use. *Plasticheskie massy,* 2001, №2, pp. 42-48.

[23] Sherieva M.L., Shustov G.B., Shetov R.A. Biodagradable compositions base on starch. *Plast.mass,* 2004, №10, pp.29-31.

[24] Legonkova OA., Peshehonova A.L., O.A.Sdobnikova. *«Tara i upackovka»,* 2003, №5, pp.56-60.

[25] Gannyshkina E.V., Krunetzkaya L.F. Survey of standards and some data on microbiological damages of plastics., Kiev, *«Naukova Dumka»,* 1978, pp.80-91.

[26] Sapozhnikova S.A. Varnish-and-paint coatings for tropical climat. Moscow, 1977. № 1. p. 4-11.

[27] Lisitzhina-Kulik E.S. Estimation of some facts on stable of varnish-and-paint coatings to damage by fungies, 1971, №4, pp.58-61.

[28] Lisitzhina-Kulik E.S., Moiseeva N.G. Varnish-and paint materials and their use, 1972, № 3, pp. 33-34.

[29] Wolfram Tanzer. Biologisch abbaubare Polymere. Deutscher Verlag fur Grunstoffindustrie, Stuttgart, 2000, 200s.

[30] Im Rahmen des 2nd International Scientific Workshop on Biodegradable Polymers and Plastics, sinngemaβe Ubersetzung nach OTTENBRITTE et al., 1992.

[31] Aminabhavi T.M., Balundgi R.H. A review on biodegradable plastics, *Polym.-Plast. Technol. Eng.,* 1990, 29(3), pp. 235-262.

[32] Dietz E. Der politische Rahmen fur alternative Verparkungen. *Vogel Buchverlag,* 1994, ss.15-22.

[33] Im Rahmen des 2nd International Scientific Workshop on Biodegradable Polymers and Plastics, sinngemaβe Ubersetzung nach OTTENBRITTE et al., 1992.

[34] Gudgabidze A.O. Investigation of mechanism of biodamage of polymers' materials in moist subtropics conditions. Tbilisi, 1990.

[35] Scrjabin G.K., Golovleva L.A. Microbiological transformation of organic combinations in co-oxydation terms. Moscow, 1972, pp.232-245.

[36] Fonkon G., Jonson V. Microbiological oxidation., Moscow, *Mir,* 1976, 239c.

[37] Alberson A.C., Ranby B. Biodegradation of Synthetic Polymers: The C-14 Method Applied to Polyethylentes. Proceedings of the Third International Biodegradable Symposium. London: *Applied Science,* 750 p.

[38] Blagodatskaya E.V. Dinamics of cellulose degradation in soil. Moscow, 1989.

[39] Zaguljaeva Z.A., Problems of biological damages and biofouling with fungi of materials, goods, constructions. Moscow, *Nauka,* 1972, pp,51-54.

[40] Vinogradova L.M., Moiseeva. Microoganisms and plants are destroyers of materials and merchandises. Moscow, *Nauka,* 1979, pp. 90-96.

[41] Ruben E.L. Microbe lipids and lipases. Moscow, 1977, 215p.

[42] Abramova N.F., Naplekova N.N. Physiological activity of fungi during their growth on plastics and their changes while storage conditions. Kiev, *Naukova Dumka,* 1978, pp.69-70.

[43] Zuev J.S. Demolition of polymers under aggressive mediums. Moscow, *Chemistry.,* 1972, 229 p.

[44] Tynnii A.N. Durability and demolition of polymers under liquid medium. Kiev, *Naykova Dumka, ,* 1975, 247 p.

[45] Manin V.N., Gromov A.N. Physical and chemical stability of polymers under exploitation terms. Leningrad, 1980, 248 p.

[46] Vorobjeva G.J. Chemical stability of polymer materials. Moscow, 1981, 296p.

[47] Moiseev J.V. Zaikov G.E. Chemical stability of polymers in aggressive mediums. Moscow, 1979, 287 p.

[48] Imanuel M.M., Buchachenko A.L. Chemical physics of aging and stabilization of polymers. Moscow, 1982, 359 p.

[49] Smirnov V.F. Investiagaion of physiological and biochemical mechanisms of action of *Aspergillius Niger* in connection with defence of polymers from biodamage. Voronezh, 1982.

[50] Abramova N.F. Biological damage of polymers. Alma-Ata, 1977.

[51] Pehtusheva E.L., Biodemage and defence of nonfoods. Moscow, 2002, 224 p.

[52] Sukhareva L.A. Gubanova M.I., Legonkova O.A. Biochemicaly stable powder composition in food industry, Moscow, 2002, 177 p.

[53] Jakovleva V.S., Bakirova E.V., Chulkov N.G. Investigation of coatings for inner surface of vessels for storage of ethyl alcohol. Moscow, 1999, pp.53-54.

[54] Sukhareva L.A. Durability of polymer coatings. Moscow, 1984, 240 p.

[55] Zubov P.I., Sukhareva L.A. Structure and properties of polymer coatings. Moscow, 1982, 256 p.
[56] Sukhareva L.A. Polyester coatings, structure and properties, Moscow, 1987, 192 p.
[57] Sukhareva L.A. Physical and chemical basis of modification of polymers' coatings. Moscow, Kotovitzhi, 1998, pp. 81-86.
[58] Sukhareva L.A. Methods of improving of properties of polymers' coatings, based on creation of well-ordered structure. Penza, 2000,.60 p.
[59] Sukhareva L.A., etc. Bactericidal rust-resistal epoxy coatings for different purposes. Penza, 1998, 81 p.
[60] Sukhareva L.A., etc. Biochemicaly stable defensive coatings for water supply. Moscow, 579 p.
[61] Sukhareva L.A., etc. Biochemicaly coatings for large capacity vessels. Penza, 2000, 45 p.
[62] Smirnov R.F. Bactericidal acryl glue compositions. Moscow, 1995.
[63] Sukhareva L.A. Jakovlev V.S. Biotechnology of defensive polymer and inorganic coatings, Moscow, 2001, 328 p.
[64] Sukhareva L.A., Jakovlev V.S., Mzhachih E.I.. Packaging materials in production and storage of food production. Moscow, 2003, 559 p.
[65] Blatnik R., Zanova V., *Microbiological rust.* Moscow, 1965, 222 p.
[66] Zaburina K.I. Electrical insulating lacquers and materials intended to be used in subtropical terms. Moscow, 1977, .№1, pp. 27-44.
[67] Corjakina M.I., Mairova N.V. Lacquers and painting materials, Moscow, 1985. N5. pp.35-37.
[68] Sukharev L.A., Semenov G.V., Sergeenko T.E.. Ecological problems of biodegradation of industrial, construction materials and industrial wastel, Penza, 1998, pp. 9-12.
[69] Sukhareva L.A., Komarov V.V. Methods of increase of defensive properties and durability of coatings based on vinyl copolymers. Moscow, 2002, 103 p.
[70] Gregory. Microbiology of atmosphere. Moscow, 1964, 291p.
[71] Mirchik S.M. Soil microbiology, Moscow, 1976.
[72] Emzhov V.T. Fundamentals of soil microbiology. Moscow, 1983, 100 p.
[73] Trisvjatzkii L.A., Grain storage. Moscow, 1986, 350 p.
[74] Zvjaginztev D.G. Soil and microorganisms. Moscow, 1987, 200 p.
[75] Zimenko T.G., Markova N.L. Microoganisms in guarding of soils from industrial contamination. Minsk, 1986, 46 p.
[76] Imranov E.L, Microoganisms in demolition of wood waste. Moscow, 2000.
[77] Legonkova O.A.,Methods of biodegradation estimation. *«Secondary resources»,* 2005, №5,pp.18-22.

In: New Topics in Monomer and Polymer Research
Editors: A. D'Amore and G. Zaikov, pp. 75-111
ISBN: 1-60021-436-3

Chapter 7

SYNTHESIS OF HETEROCYCLIC ORGANOSILICON DI- AND POLYFUNCTIONAL COMPOUNDS

O. Mukbaniani [1]***, G. Zaikov*** [2]***,***
T. Tatrishvili [1] ***and N. Mukbaniani***

[1] I. Javakhishvili Tbilisi State University, I. Chavchavadze Ave., 1, 0128 Tbilisi, Georgia
2 Institute of Biochemical Physics Russian Academy of Sciences 4 Kosygin Street, Moscow 119991, Russia

INTRODUCTION

The present state of the synthesis of cyclic organosiloxanes and organocarbosiloxanes with assigned disposition of various functional groups is reviewed.

Organosilicon monomers containing functional groups at silicon atoms are extremely inclined to various reactions of transformation, which produce ≡Si-O-Si≡ bonds [1]. At the present time, the most practically significant organosilicon polymers are industrially produced from organochlorosilanes of various functionalities [2]. To carry out directed synthesis of oligomers and block-copolymers with cyclic and polycyclic fragments in the macromolecular backbone, methods of linear di- and poly-functional oligochlororganosiloxanes synthesis have been developed [3, 4], on the basis of which or-ganocyclosilanes with the given disposition of chlorine atoms and other functional groups at atoms of silicon may be derived.

1. ORGANOCYCLOSILOXANES WITH GIVEN DISPOSITION OF FUNCTIONAL GROUPS AT ATOMS OF SILICON

Synthesis of organocyclosiloxanes with different disposition of functional groups at atoms of silicon is mainly reduced to two methods: hydrolytic condensation of chlorine-containing organosilanes and alkoxyorganosilanes and HFC reaction of silanes with siloxanes containing functional chlorine-, alko-xy-, amino and hydroxyl groups or to modification of previously prepared organocyclosiloxanes.

Hydrolytic condensation of dibutoxydichlorosilane in the presence of hydrogen chloride acceptor (tertiary amine) was used for synthesizing organocyclosiloxanes with butoxyl groups [5, 6]. Similarly, hydrolytic condensation of diethoxydichlorosilane proceeds in the presence of pyridine [7]. Partial hydrolytic condensation of tetraethoxysilane and hexaethoxydisiloxane [8,9] induces formation of organocyclosiloxanes with ethoxy groups at silicon atoms.

Hydrolytic condensation of alkylalkoxydichlorosilanes in the presence of pyridine also produces organocyclosiloxanes containing, besides alkoxy groups, alkyl ones at silicon atoms [10, 11].

Joint reactions of $SiCl_4$ etherification and hydrolytic condensation with isopropyl alcohol produce organocyclosiloxanes with isopropyloxy groups at silicon atoms [12].

Joint hydrolytic condensation between dimethyldichlorosilane and methylalkoxydichlorosilanes in the presence of pyridine has produced organocyclosiloxanes with two alkoxy groups at different silicon atoms [13]:

$$2Me_2SiCl_2 + 2MeSi(OR)Cl_2 + 4H_2O + 8Py \longrightarrow [Me_2SiO][MeSi(OR)O]_2 + 8Py \cdot HCl$$

Scheme 1.

Hydrolytic co-condensation between diphenyldichlorosilane and phenyltrichlorosilane at 1:4 molar ratios of the initial components produces hydroxypentaphenylcyclotrisiloxane and 1,5-dihydroxy-he-xaphenylcyclotetrasiloxane [14]:

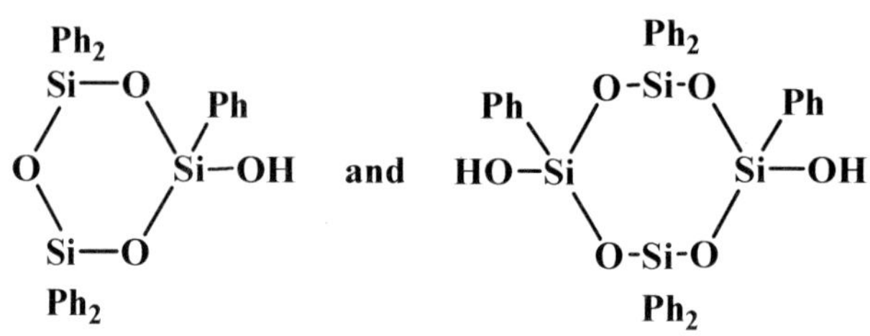

Scheme 2.

However, it should be noted that the yield of individual phenylcyclosiloxanes is negligible.

Joint hydrolysis of dimethyldichlorosilane and methylvinyldichlorosilane produced divinylhexame-thylcyclotetrasiloxane [15]. Despite the application of efficient rectification towers as well as ana-lytical chromatograph with preparative attachment, the authors have not managed to separate isomeric 1,3- and 1,5-divinylhexame-thylcyclotetrasiloxanes, which may be produced by joint hydrolysis. The reaction proceeds in accordance with the general scheme as follows:

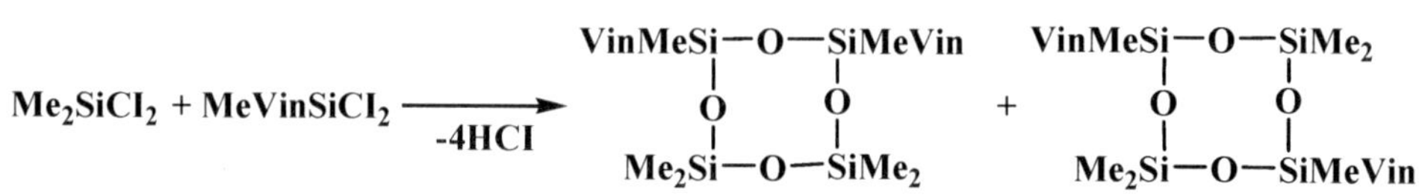

Scheme 3.

In accordance with the scheme as follows [16], 1-hydride-3-vinylhexamethylcyclotetrasiloxane was synthesized in co-hydrolysis reaction between 1,3-dichlorotetramethyldisiloxane and 1-chloro-3-chlo-ro-3-vinyltetramethyldisiloxane:

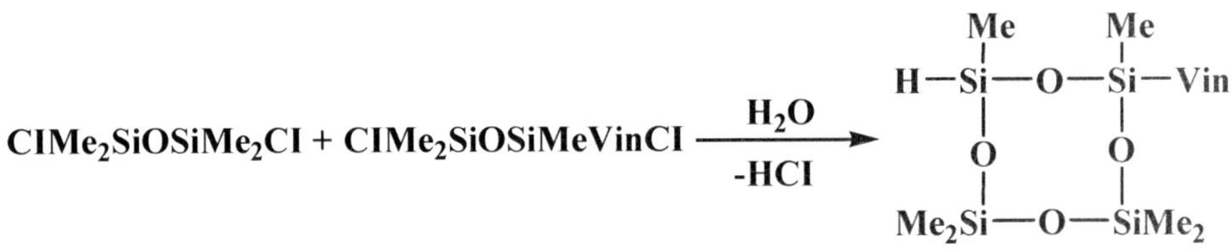

Scheme 4.

Hydrolytic condensation of phenyltrichlorosilane in aqueous solution of acetone, in acidic medium has synthesized *cis*-1,3,5,7-tetrahydroxy-1,3,5,7-tetraphenylcyclotetrasiloxane (tetrole) with high yield [17]. The reaction proceeds by the scheme shown below:

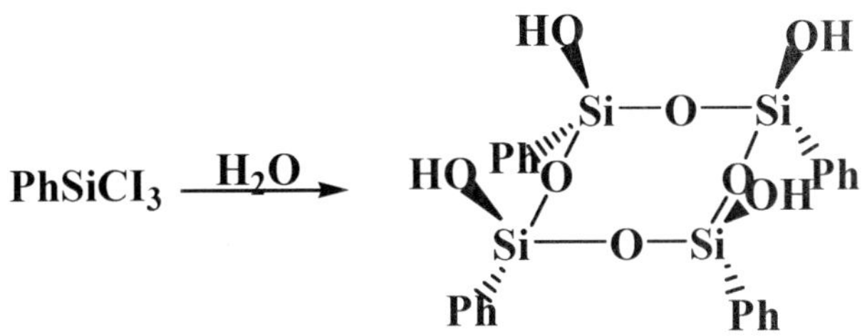

Scheme 5.

Later on, mutual configurative transformations of four *stereo*-isomers of tetrole were detected. These stereo isomers are of the following structures:

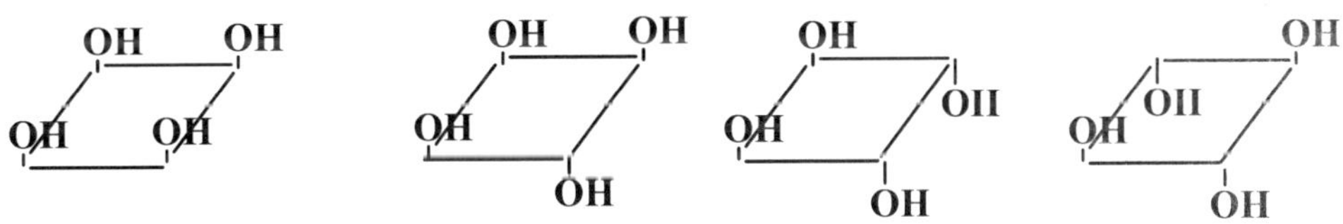

Scheme 6.

It has been shown [18] that isomerization of tetrasiloxane ring may proceed both with reversal of silicon atoms configuration and without it.

Tetrole-like compounds were also synthesized by hydrolytic condensation of (β-phenyl)phenethyl-trichlorosilane, stilbene trichlorosilane and (α-phenyl) phenethyltrichlorosilane [19].

At 20-50°C in acetone, partial hydrolytic condensation of 1,1,3,3-tetrachloro-1,3-diphenyldisiloxane [20] proceeds with 20% yield of cyclotetraandcyclohexasiloxane products, $[PhClSiO]_4$ and $[PhClSiO]_6$. Low yield of cyclic products is explained by formation of a significant quantity of linear products.

Besides 1,5-dimethyl-1,5-divinyl-3,3,7,7-tetraphenylcyclotetrasiloxane, the interaction between equi-molar amounts of diphenylsilanediol and methylvinyldichlorosilane also synthesizes 1,3-dimethyl-1,3-divinyl-5,5-diphenylcyclotrisiloxane, 1-methyl-1-vinyl-3,3,5,5-tetraphenyltrisiloxane and high-molecular products in accordance with the scheme as follows [21]:

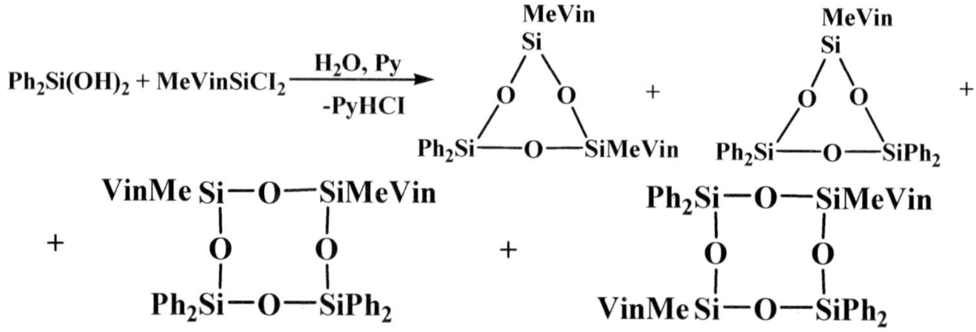

Scheme 7.

HFC reaction is more suitable for synthesizing organocyclosiloxanes with functional groups at silicon atoms. High yields of organocyclosiloxanes with a single functional group are produced in double re-placement reaction of disodium salts of dimethylsiloxanes with ethylbutoxydichlorosilane [14], me-thylbutoxydichlorosilane [22] and methyltrichlorosilane [23].

The HFC reaction between 1,1,3,3-tetrachlorodiphenyldisiloxane and dihydroxydiorganosilanes in the presence of hydrogen chloride acceptor causes formation of 1,3-dichlorotetraorganocyclotrisiloxanes [24] in accordance with the scheme as follows:

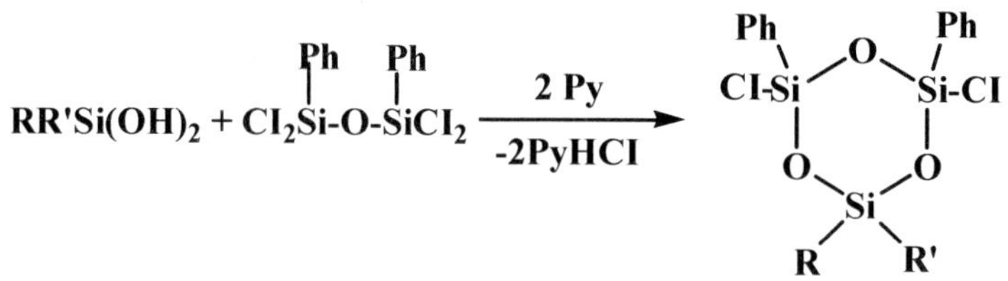

Scheme 8.

Where: R = R' = Ph, p-$CH_3C_6H_5$; R $\neq$ R'; R = Ph, R' = Me.

Hydrolysis of synthesized 1,3-dichlorotetraorganocyclotrisiloxanes has produced appropriate dihyd-roxy derivatives.

Condensation reactions of dihydroxydiphenylsilane or dihydroxydimethylsilane with methylvinyl-dichlorosilane and methyldichlorosilane [25, 26] in the presence of pyridine (1:1 molar ratio of the initial components) have given high yields of organocyclotetrasiloxanes with 1,5-disposition of func-tional groups at silicon atoms. The reaction proceeds as follows:

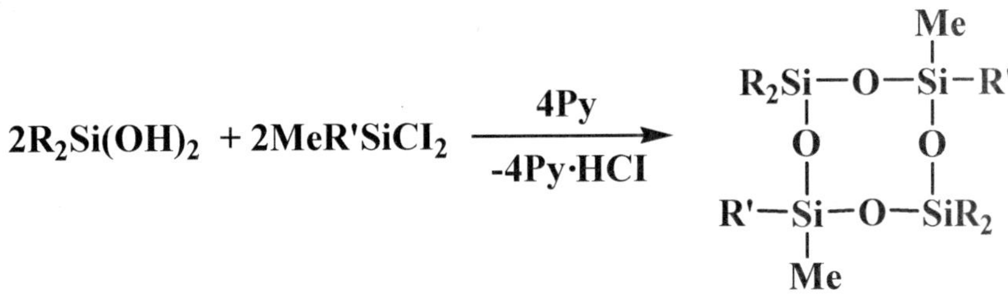

Scheme 9.

Where: R = Me, Ph, Et; R' = Vin, H.

The study of HFC reaction of 1,3-dihydroxytetraphenylsiloxane with silicon tetrachloride at equi-molar ratio of the initial components has indicated [27] that besides organocyclosiloxanes, linear polyfunctional organosiloxane are also produced in accordance with the scheme as follows:

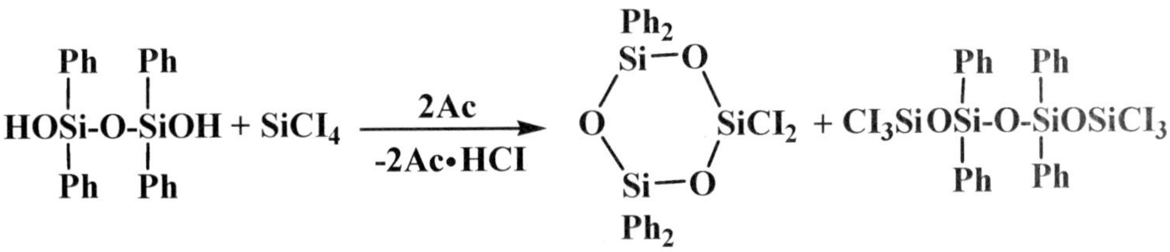

Scheme 10.

The yield of linear products is increased with the quantity of silicon tetrachloride. In a similar manner, equimolar HFC reaction of silicon tetrachloride with 1,3-dihydroxy-1,3-divinyl-1,3-diphenyldisiloxa-ne has been studied [27].

The HFC reaction of 1,3-dihydroxytetraphenyldisiloxane with 1,5-dihydroxyhexaphenyltrisiloxane and organochlorosilanes at 1:3 ratio of initial components in anhydrous toluene solution at -5°C in the presence of pyridine under the above-mentioned conditions proceeds ambiguously and besides linear organochlorosiloxanes, the yield of which equals 70 – 87%, organocyclotrisiloxanes and organocyc-lotetrasiloxane are also produced in low amounts (4 – 6%). As a consequence, the reaction proceeds in accordance with the general scheme as follows [28, 29]:

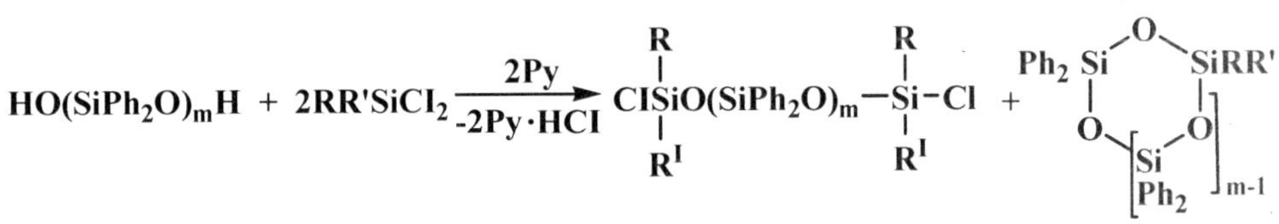

Scheme 11.

Where: $m = 2, 3$; R = Ph, Me, Vin, Cl; R' = Cl, H, Vin.

The HFC reaction between methyldichlorosilane, dihydroxydimethylsilane and 1,3-dihydroxytetra-methyldisiloxane at 1:1 molar ratio of initial components in the presence of acceptor (amine) mainly proceeds with formation of linear tetrachlororganosiloxanes [30]:

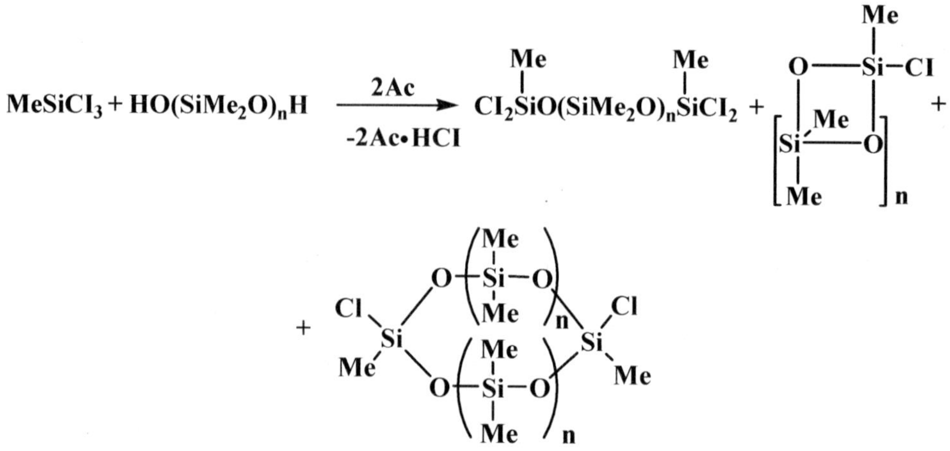

Scheme 12.

Where: $n = 1, 2, 3$.

The same investigators [31] have also obtained linear tetrachloroethylsiloxanes and cyclic ethyl-siloxanes by the HFC reaction of ethyltrichlorosilane, dihydroxydiethylsilane with 1,3-dihydroxytetra-ethyldisiloxane in chlorosilane excess.

The HFC reaction of dihydroxydiorganosilanes disodium salts with organochlorosilanes causes for-mation of linear, cyclic and star-shaped siloxanes [32 – 35].

Investigation of the HFC reaction between 1,3-dihydroxy-1,3-dimethyl-1,3-diphenyldisiloxane and 1,5-disodiumoxyhexamethyltrisiloxane and silicon tetrachloride, methyl- and phenyltrichlorosilanes has displayed that with regard to the ratio of initial components both linear and cyclic compounds may be produced. The yield of linear reaction products is increased with excess of chlorosilanes in the primary reaction mixture, and the yield of cyclic compounds is decreased.

Divinyl- and dihydride-containing organocyclosiloxanes with different disposition of functional groups at silicon atoms were synthesized by the HFC reaction of 1,3-dihydroxyorganodisiloxanes with dichlorodiorganosilanes and 1,3-dichlororganodisiloxanes in the presence of pyridine in accor-dance with the general scheme as follows [25, 36]:

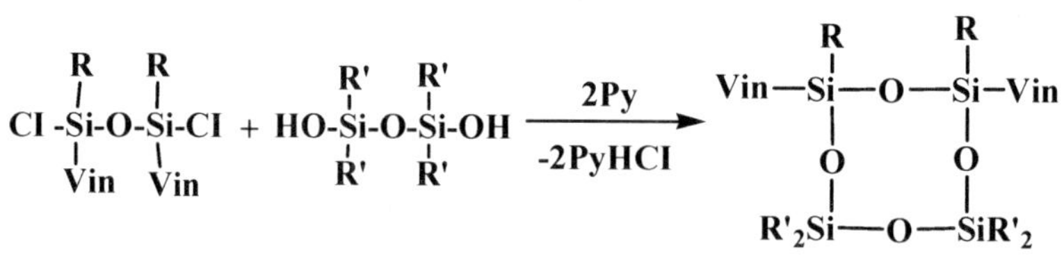

Scheme 13.

Where: R = R' = Me, Ph; R ≠ R'.

By analogy, heterofunctional condensation of dihydroxydiphenylsilane with methyltrichlorosilane [36, 37], silicon tetrachloride [38] and phenyltrichlorosilane [39, 40], as well as between diethyl-dihydroxysilane and ethylvinyldichlorosilane and ethyldichlorosilane, proceeds at equimolar (1:1) ra-tio of the initial components, in the presence of pyridine. The yield of target products with 1,5-dis-position of chlorine atoms at silicon ones is varied from 41 to 55%. The reaction scheme is as follows:

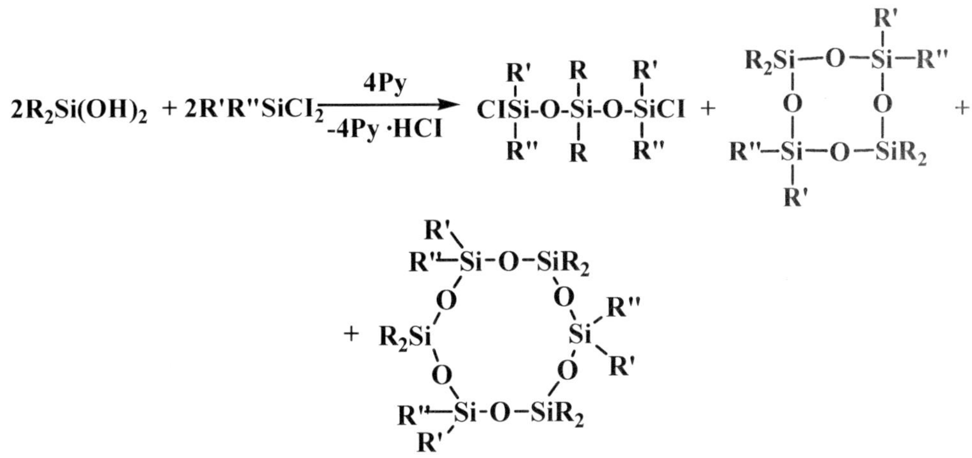

Scheme 14.

Where: R=Me, Ph, Et; R'=Me, Ph, Et, CI, Vin; R"=CI, H, Vin.

As shown in the above scheme, besides isomeric organocyclosiloxanes, linear tetrachlororganotrisi-loxanes are also synthesized with the yield of 15 –20%. For methyl-containing organocyclotetra-siloxanes, two singlet signals with almost equal peak ratio (53 – 47%) at 0.38 and 0.42 ppm. have been detected in ^{1}H NMR spectra. These ranges are typical of *cis*- and *trans*-forms of methyl groups. Shift for methyl protons to a weak field testifies about the electron acceptor type of functional groups at the same silicon atom, where methyl groups are disposed. To increase the yield of organocyclosilo-xanes with 1,5-disposition of chlorine atoms at silicon ones, the two-stage synthesis method was suggested [41, 38, 40]. At the first stage, HFC of dihydroxydiphenylsilane with excess of organotri-chlorosilane is carried out. At the second stage, HFC of obtained tetrachloroorganotrisiloxanes with dihydroxydiphenylsilane proceeds. The scheme of the reaction is as follows:

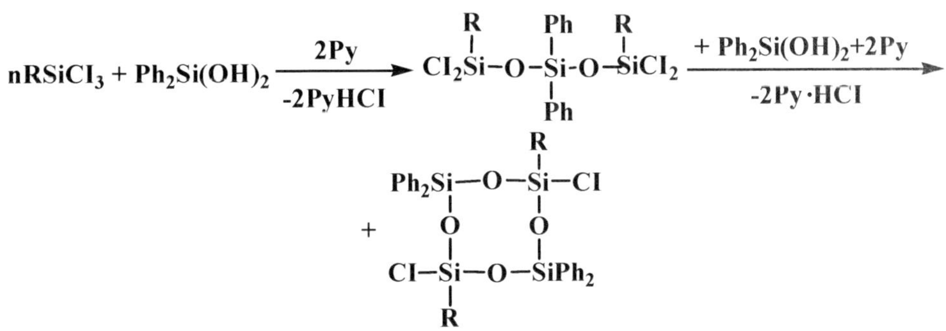

Scheme 15.

Where: R = Me, Ph, CI.

It is shown [42] that depending on the ratio of initial reagents,diorganodiaminosilanes mixed with diorganodichlorosilanes interact with diphenylsilanediol giving cyclic or linear siloxanes. The yield of mixed cyclotetrasiloxanes is quantitative and almost independent of the type of framing at silicon atoms in amino- and chlorosilanes. Trisiloxane is synthesized with lower yield (~70%) due to a noti-ceable formation of secondary products, which are long-chain α,ω−dichlorosiloxanes.

Formation of long-chain siloxanes is much more typical of the condensation reaction proceeding with dihydroxydiphenylsilane excess. In this case, the yield of trisiloxane does not exceed 50%. The reac-tion proceeds by the schemes as follows:

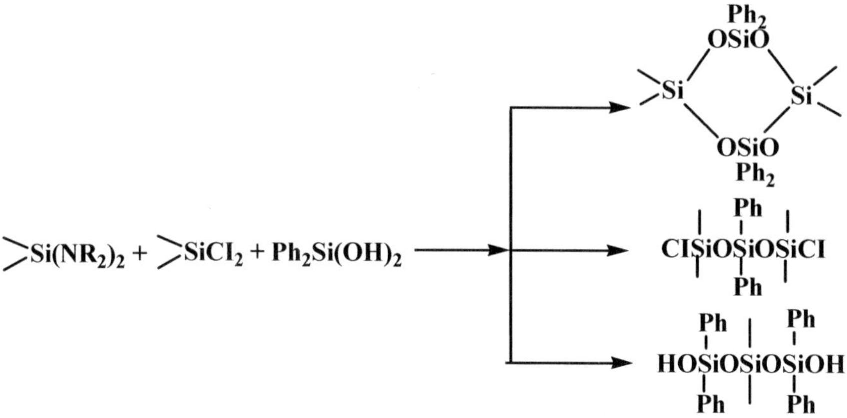

Scheme 16.

For obtaining high yields of synthesized methylcyclosiloxanes with given disposition of functional groups at silicon atoms, the two-stage synthesis was suggested [31, 43, 44] in accordance with the scheme as follows:

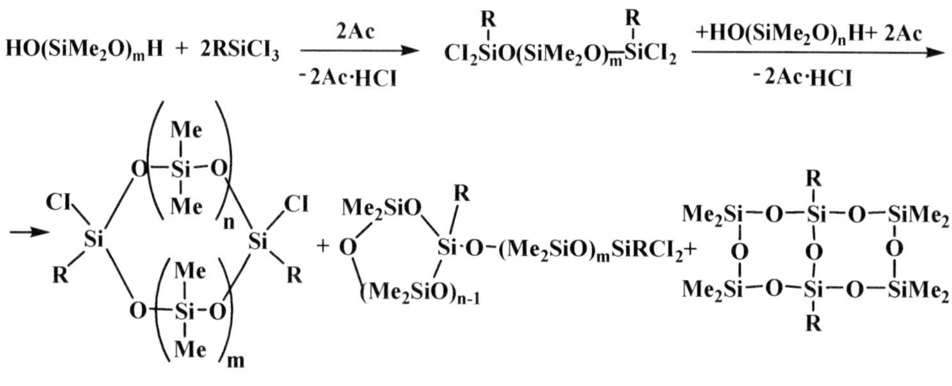

Scheme 17.

Where: m = n = 1, 2, 3; m ≠ n; R = Me, Ph.

Two-stage reaction proceeding resulted in the yield of organocyclotetra(penta, hexa)siloxanes equal 28 – 51%. The ratio of dichlorocyclosiloxane isomers was monitored by ESR and mass-spectrometry methods. Basically, dichlororganocyclosiloxanes of 96.0–99.5% purity were produced. An attempt to separate dichlororganocyclotetra(penta, hexa)siloxanes to stereo-isomers was made. The majority of dichlororganocyclotetra(hexa)siloxanes are separated to *cis*- and *trans*-isomers. All attempts to sepa-rate stereo-isomers of dichloroctaorganocyclopentasiloxane and dichlorododecaorganocycloheptasi-loxane to *cis*- and *trans*-isomers have failed. Hydrolysis and aminolysis of dichloromethylcyclotetra-(penta, hexa, hepta, octa)siloxanes proceeding in the presence of hydrogen chloride acceptors have produced corresponded dihydroxy- and diaminomethylcyclosiloxanes. Inversion of isomers configu-ration during hydrolysis was Separation methylcyclosiloxane dihydroxy-derivatives to *cis*- and *trans*-isomers was performed for 1,5-dihydroxyhexamethylcyclotetrasiloxane and 1,7-dihydroxydeca-methylcyclohexasiloxane by recrystallization from heptane; under similar conditions, the rest hydro-xyderivatives cannot be separated. X-ray diffraction studies of the crystalline structure of 1,5-dihyd-roxy-decamethylcyclohexasiloxane *cis*- and *trans*-isomers show that ≡Si-O-Si≡ bond angle in *cis*-isomers displays two values: 145.7 – 150.2° and 173.6 –174.1°. The ring is tub-shaped with silicon atoms in the hydroxyl group distant from the middle atom plane by 0.96-0.98 Å. This allows OH-groups in *cis*-isomers forming intra- and intermolecular hydrogen bonds [45].

In the range of 3200 – 3800 cm^{-1} typical of valence oscillations of OH groups, IR spectra of dihyd-roxymethylcyclo(tetra, hexa)siloxanes, mould with KBr and in CCl_4 solutions of various concentra-tion display the absence of a broad absorption band at 3300 – 3500 cm^{-1}, which is present in spectra of all dihydroxymethylcyclosiloxanes mould with KBr, and the presence of a narrow band at 3700 cm^{-1} typical of hydroxyl group with intermolecular hydrogen bonds, which confirms separation of pure *trans*-isomer of dihydroxymethylcyclosiloxanes. For *cis*-isomers, IR spectra of diluted solution po-ssess two bands: a broad one in the range of 3300 - 3500 cm^{-1} typical of intramolecular hydrogen bond and a narrow one at 3700 cm^{-1}.

Based on ^1H NMR spectra, chemical shifts and multiplicity of methyl protons of chloro(hydroxy)-substituted methylcyclotetra(penta, hexa)siloxanes were studied [46].

By similar two-stage HFC method, difunctional methyl(ethyl)cyclotetra(penta, hexa)siloxanes with defined disposition of chlorine, hydride or vinyl groups at silicon atoms were produced [47, 48].

HFC reaction of organodichlorosilanes or organotrichlorosilane with 1,3-dihydroxytetraphenyldisilo-xane under usual conditions (equimolar (1:1) ration of initial reagents, presence of hydrogen chloride acceptor – pyridine, diluted toluene solution, 0 - 5°C temperature range) mainly produces both lincar organochlorosiloxanes and cyclic compounds. The presence of linear organochlorosiloxanes, which content in the reaction mixture did not exceed 6%, was confirmed by GLC analysis. The reaction pro-ceeds as follows [49,50]:

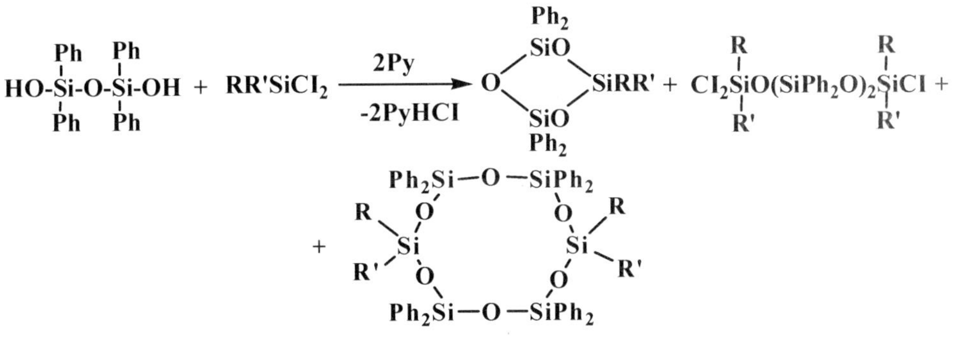

Scheme 18.

Where: R = Me, Ph, CI; R'= CI, Vin, H.

To the authors' point of view, similar to the case described in ref. [51], the reaction proceeds with for-mation of a transient formation:

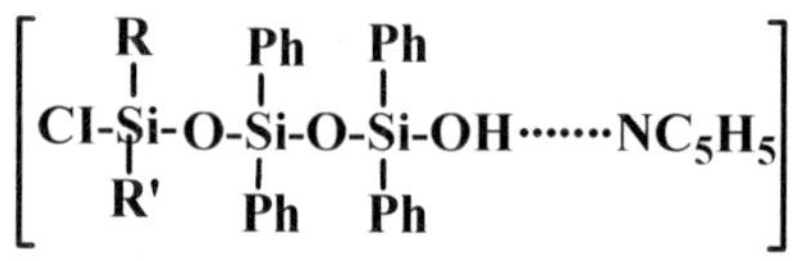

Scheme 19.

which then is subject to intramolecular condensation producing a hexatomic rings and intermolecular condensation with further cyclization of condensation products. The latter leads to formation of dode-catomic rings and higher boiling products. The study of NMR spectrum of 1,7-dichloro-1,7-dime-thyloctapenylcyclohexasiloxane has indicated that HFC reaction of 1,3-dihydroxytetraphenyl-disilo-xane produces cyclohexasiloxane and its structural isomer:

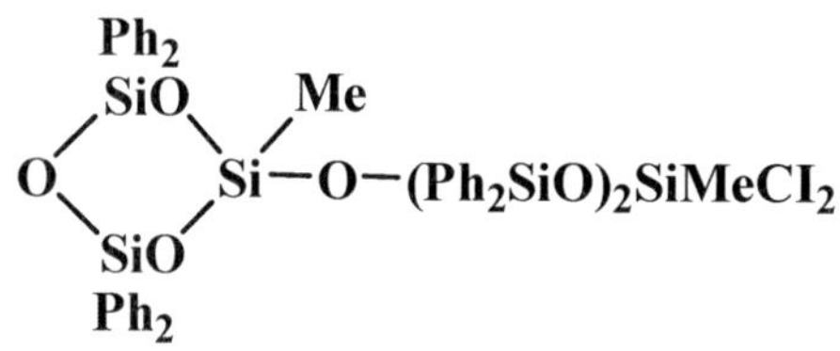

Scheme 20.

A singlet at 0.27 ppm is corresponded to methyl group of methyldichlorosiloxy-fragment; for methyl protons in silsesquioxane unit, a singlet at 0.46 ppm is also observed. Two singlets at 0.38 and 0.41 ppm, corresponded to two methyl groups of *cis*- and *trans*-forms, respectively, are observed in spec-tra. After recrystallization from toluene-heptane mixture, organocyclohexasiloxane product was puri-fied from isomeric admixtures. The ratio of *cis*- and *trans*-isomers equaled 54 and 46%, respectively. Accordingly, after fractionation, HFC reaction products of organotrichlorosilanes and silicon tetra-chloride with 1,3-dihydroxytetraphenyldisiloxane were re-crystallized from the toluene-heptane mix-ture. HFC reaction of dichlorodiorganosiloxanes with 1,3-dihydroxytetraphenyldisiloxane did not pro-duce isomeric products. At the single-stage synthesis, the yield of organocyclohexasiloxanes equaled 26 – 43%. HFC reaction of tetrachloro- or dichlororganotetrasiloxanes with dihydroxyophenylsilane or 1,3-dihydroxytetraphenyldisiloxane under usual conditions (at equimolar ratio of initial com-ponents, in the presence of pyridine, in diluted anhydrous toluene solution) produced organocyc-lopenta- and organocyclohexasiloxanes with 1,5- and 1,7-disposition of chlorine atoms at silicon. The reaction scheme is as follows [52 – 55]:

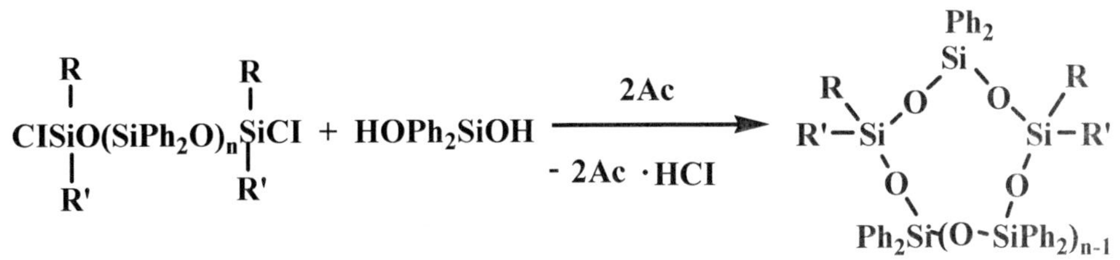

Scheme 21.

Where: n = 1, 2; R = Me, Ph; R'= CI, H, Vin.

By HFC reaction of hexa(tetra, di)chlororganotetra(penta)siloxanes with dihydroxydiphenylsilane un-der usual conditions (in the presence of pyridine, at equimolar ratio of initial reagents, in diluted an-hydrous toluene solution, in -5 -0°C temperature range), high yields of difunctional organocyclopen-tasiloxanes and organocyclohexasiloxanes with 1,5-disposition of functional groups were obtained. Therefore, the yield of organocyclopenta(hexa)siloxanes is increased to 61 – 70%. General reaction scheme is as follows [53, 56, 57]:

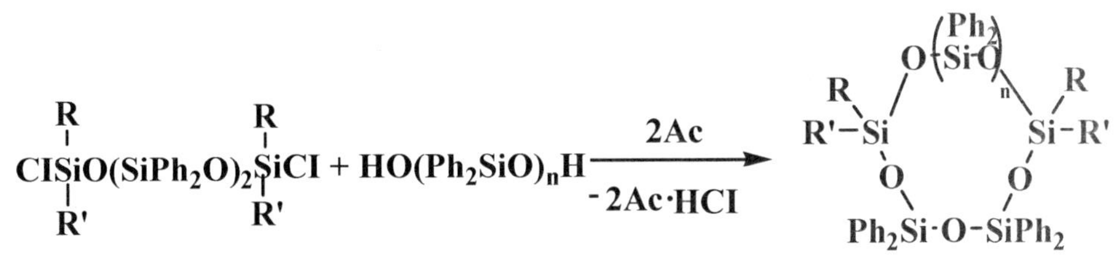

Scheme 22.

Where: n = 2, 3; R = Me, Ph, CI; R' = CI, H, Vin.

Hydrolysis of tetrachlororganocyclotetra(penta, hexa)siloxane proceeding in the presence of hydrogen chloride acceptor is described in the literature [41, 29, 58 – 61]. It is found that hydrolysis proceeding at room temperature produces polyfunctional oligomers, which are transformed into structured pro-ducts. However, as the reaction temperature was decreased to 0 -5°C, in the presence of sodium bicar-bonate, tetrahydroxy-derivatives were synthesized by the following scheme:

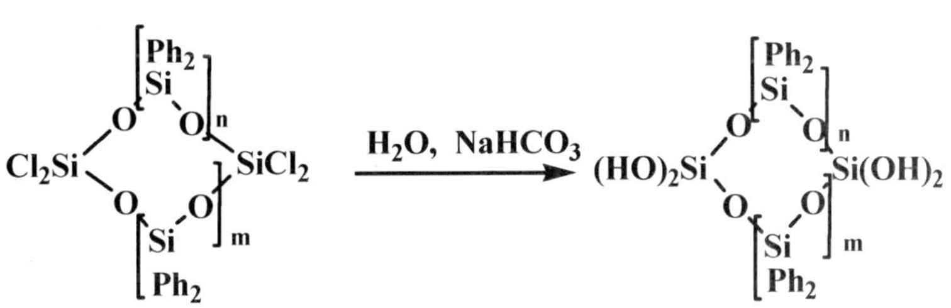

Scheme 23.

It is also found that 1,1,5,5-tetrahydroxytetraphenylcyclotetrasiloxane forms a stable structure with pyridine [59]:

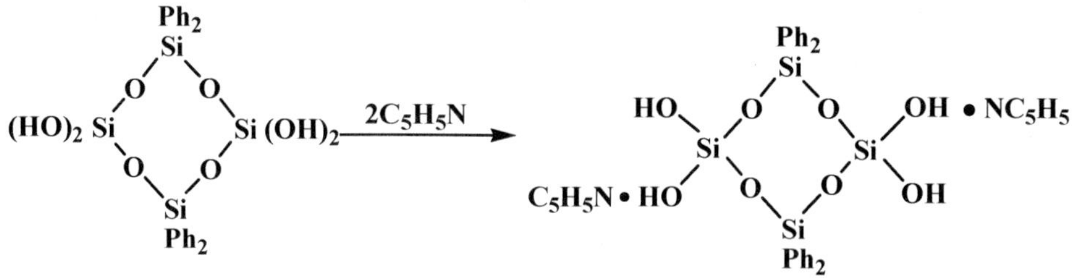

Scheme 24.

that was not observed for its isomer, cis-1,3,5,7-tetrahydroxytetraphenylcyclotetrasiloxane(tetrole).

It is assumed [62] that the structure formation is associated with increased acidity of hydrogen in hydroxyl groups, caused by interaction of *p*-electrons of oxygen atoms with silicon. Inertness of the second hydroxyl group to pyridine is explained by decrease of its acidity as a consequence of redis-tribution of electron density in the group:

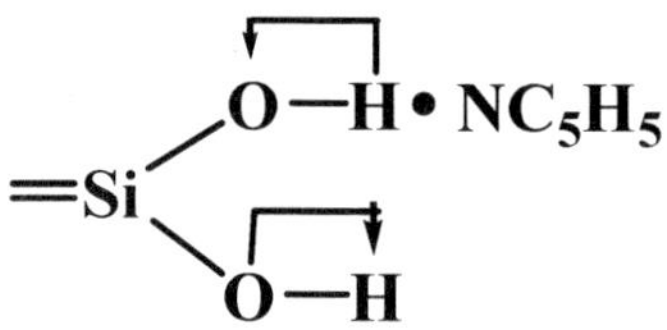

Scheme 25.

Dipyridine-1,1,5,5-tetrahydroxytetraphenylcyclotetrasiloxane structure enters HFC reaction with dior-ganodichlorosilanes or diorganoethoxychlorosilanes and forms corresponded tetrafunctional compo-unds by the following scheme:

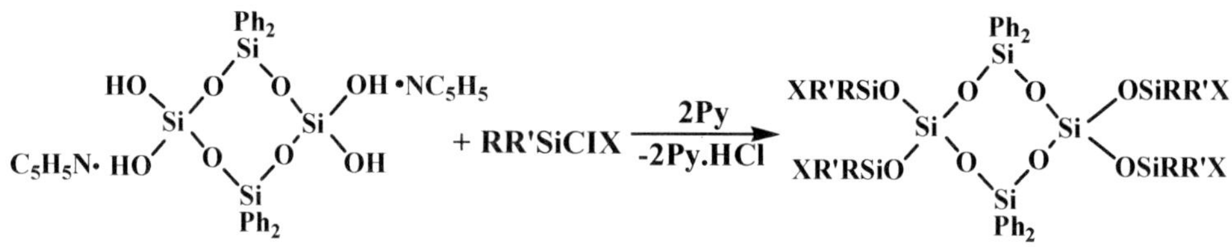

Scheme 26.

Where: R = Me, R' = Me, Ph; X = CI, OEt.

The above-mentioned reaction produces quite pure products and, consequently, their production does not require repeated fractionation, which is necessary for the case of HFC of tetrahydroxytetraphenyl-cyclotetrasiloxane with diorganodichloro- or diorganoethoxy-chlorosilanes [62].

Hydrolysis of tetrachlorine-containing branched cyclic compounds with methyl and phenyl substitu-ting agents at silicon atoms produced corresponded tetrahydroxy-derivatives [62, 64]:

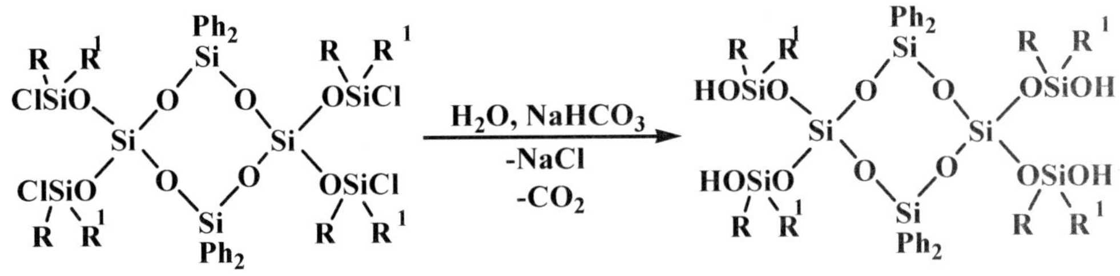

Scheme 27.

Where: $R = R_1 = Me, Ph; R \neq R_1$.

Hydrolysis proceeds with the water excess at cooling, in the presence of hydrogen chloride acceptor (R = Me) or at room temperature without it (R = Ph).

HFC reaction of 1,1,3,3-tetrachlorodiphenyldisiloxane with 1,3-dihydroxytetraphenyldisiloxane at 1:1 molar ratio of initial components in the presence of pyridine synthesizes organocyclotetrasiloxane with 1,3-disposition of chlorine atoms at silicon with ~50% yield, as well as a mixture of extended isomeric-structured rings. The reaction mainly proceeds in accordance with the scheme as follows [37,65]:

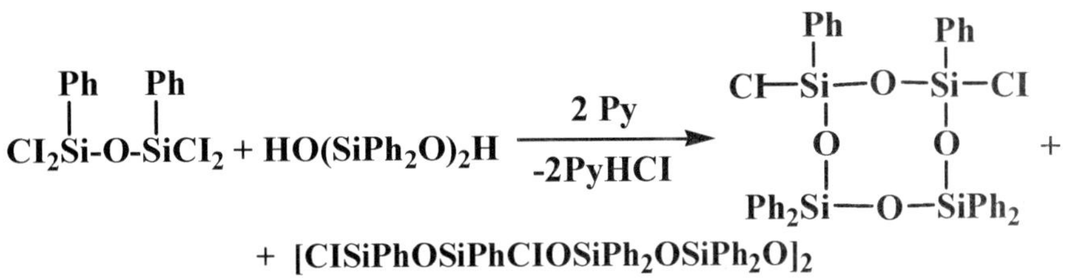

Scheme 28.

Hydrolysis of 1,3-dichlorohexaphenylcyclotetrasiloxane has synthesized an appropriate dihydroxy-derivative.

HFC reactions of 1,3-dihydroxytetraorganodi- and 1,5-dihydroxyhexaorganotrisiloxanes with silicon tetrachloride in the presence of amine or hydrogen chloride blowing off the reaction mixture have syn-thesized 1,1-dichlorotetraorganocyclotri- and 1,1-dichlorohexaorganocyclotetrasiloxanes in accordan-ce with the scheme as follows [63, 66]:

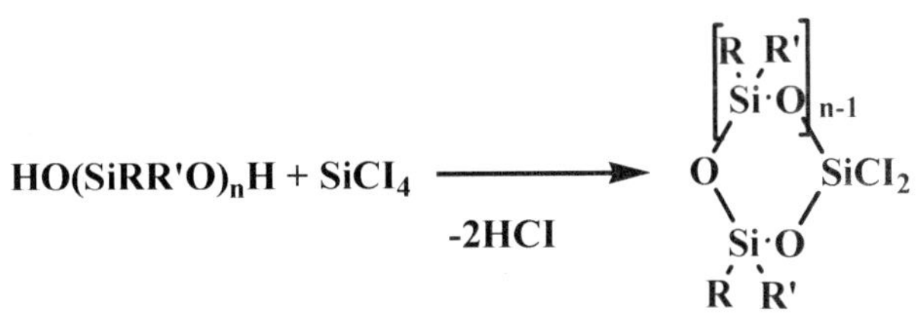

Scheme 29.

Where: $n = 1, 2$; $R = R' = Me, Ph; R \neq R'$.

The yield of cyclotrisiloxanes is decreased with the size of framing groups at silicon atoms in the initial disiloxane. Application of an acceptor gives better results compared with hydrogen chloride blowing, because in this case the secondary reactions are noticeably eliminated and the yield of target cyclosiloxanes is increased. It has been shown that affinity of 1,3-dihydroxytetraorganodisiloxanes to homocondensation in HFC reactions increases with the decrease of the framing group size at silicon atoms, which results in reduction of the yield of hex atomic organochlorocyclosiloxanes. For example, if at R=R'=Ph the yield of 1,1-dichlorotetraphenylcyclotrisiloxane equals 62%, then at R=Me and R'=Ph the yield of the appropriate cycle does not exceed 10%.

Hydrolysis of synthesized dichlororganocyclosiloxanes produces appropriate dihydroxy-derivatives. However, it has been found that hydrolysis of dichlororganocyclosiloxanes is not limited by formation of dihydroxyorganocyclosiloxanes, because besides the main reaction, intermolecular condensation of the target product with formation of more complex cyclic structures does also proceed [67].

It has been shown [68, 69] that 1,1-dichlorotetraorganocyclotrisiloxanes easily enter reactions of chlo-rine substitution by alkylamino groups:

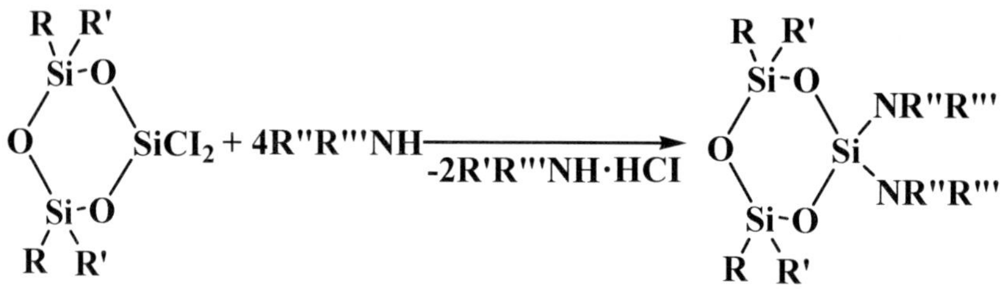

Scheme 30.

Where: R = Me, Et; R" = R"'= H, Me, Et; R" ≠ R"'.

The HFC reaction of 1,1-dichloorganocyclosiloxanes with 1,3-dihydroxytetraorganosiloxane at 1:1 ratio of initial components in the presence of pyridine has synthesized spirocyclosiloxanes of the fol-lowing structure [70, 71]:

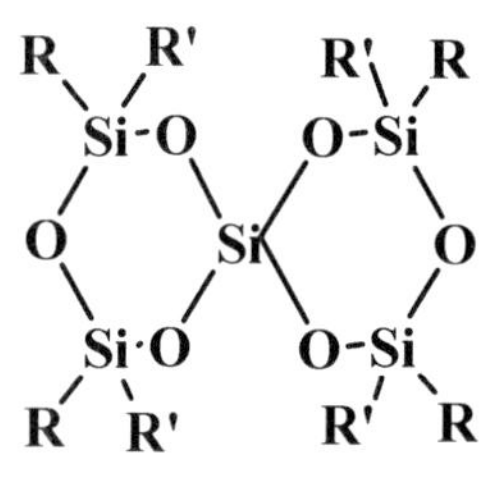

Scheme 31.

Where: R = R' = Me, Ph; R ≠ R'.

Further on, synthesized spirocyclosiloxanes were used for synthesizing cyclolinear polymers. Some physical and chemical parameters of synthesized organocyclotri(tetra, penta, hexa)siloxanes with different framing and functional groups are shown in Tables 1 – 6.

Table 1. Physical and chemical parameters of difunctional organophenylcyclotri (tetra, penta, hexa)siloxanes

№	Compound	Yield %	T_{boil}, ^{0}C (P, mmHg)	T_{melt}, ^{0}C	^{1}H(Me) chemical shift		Reference
					cis	*trans*	
1	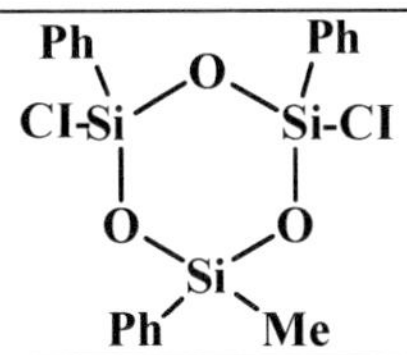	26	136-137 (0,1)	–	–	–	[24]
2	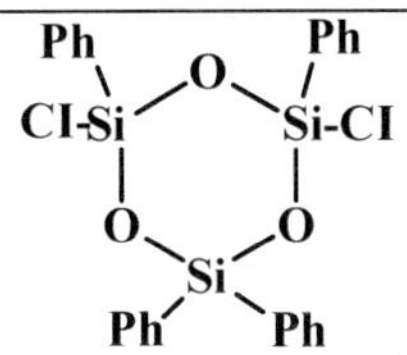	29	200-203 (0,001)	–	–	–	[24]
3	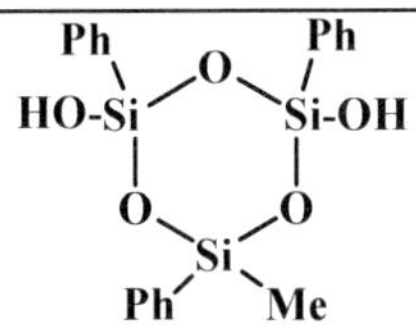	46	–	–	–	–	[24]
4	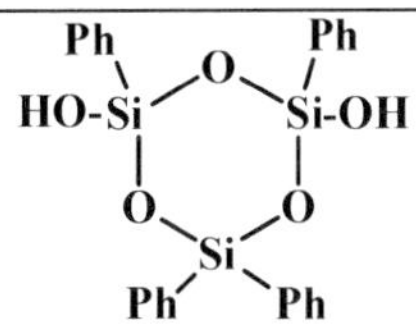	47	–	–	–	–	[24]
5	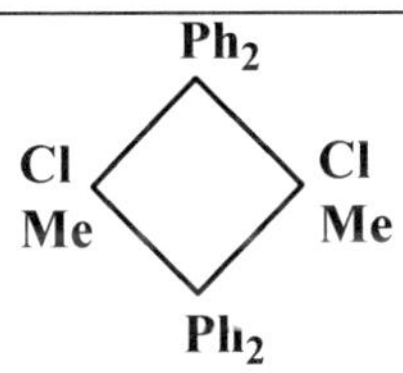	20 (66)*	247-250 (0,02)	–	0,38	0,42	[36,37]
6	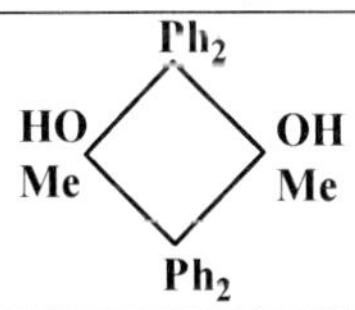	70	–	*trans* 141-142 *cis* 147-150	0,16	0,09	[37]
7	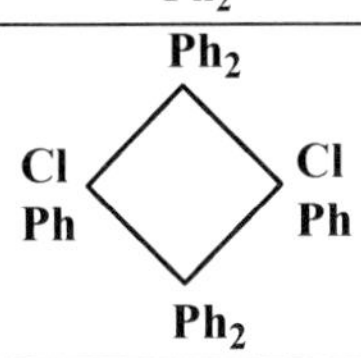	34,0 (65)	305-310 (0,02)	–	–	–	[37,65]
8	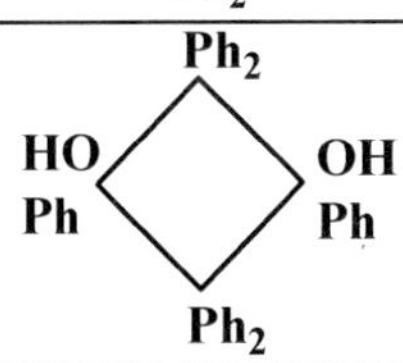	77	–	(mixture)	–	–	[37]

Table 1. (Continued)

№	Compound	Yield %	T_{boil}, ^{0}C (P, mmHg)	T_{melt}, ^{0}C	^{1}H(Me) chemical shift		Reference
					cis	*trans*	
9	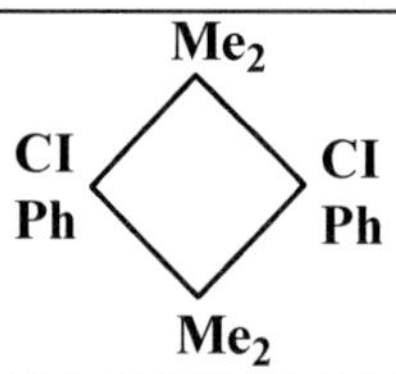	60	147-148 (0,5)	*trans* 116-117	0,42	0,29 0,16	[44]
10	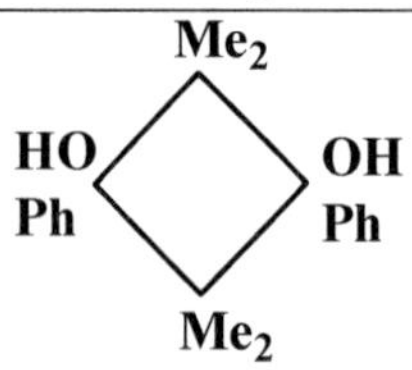	70	–	*trans* 126-127	–	0,12	[44]
11	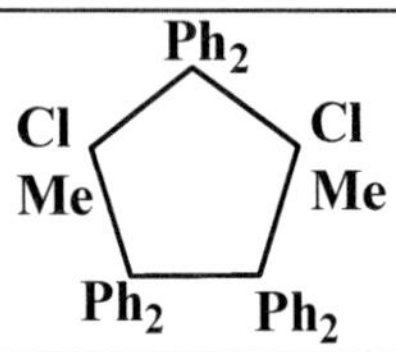	48 (79)	222-224 (760)	(mixture)	0,51	0,55	[50,53]
12	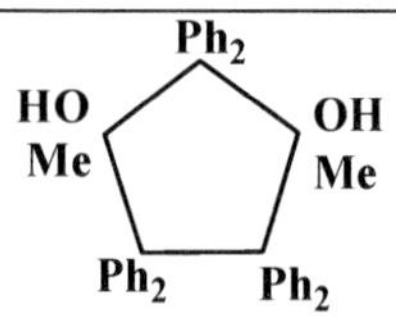	68	–	(mixture) 67-70	–	–	[50,53]
13	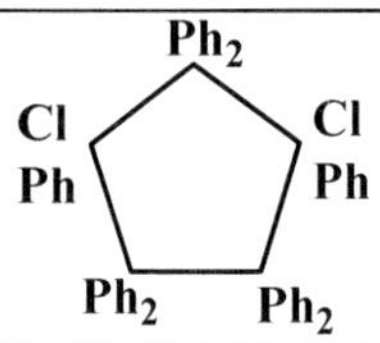	62	310-320 (0,04)	–	–	–	[50,53]
14	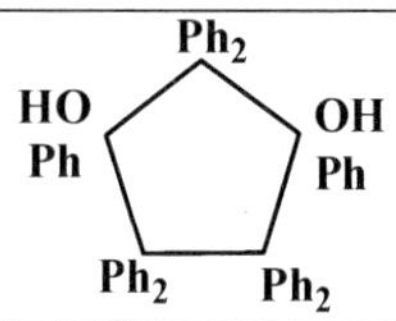	73	–	(mixture) 84-85	–	–	[50,53]
15	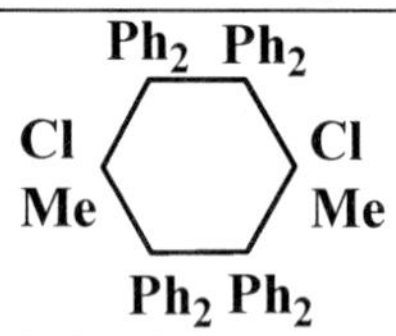	43 (61)	325-330 (0,01)	(mixture) 59-61	0,38	0,41	[49,50,52]
16	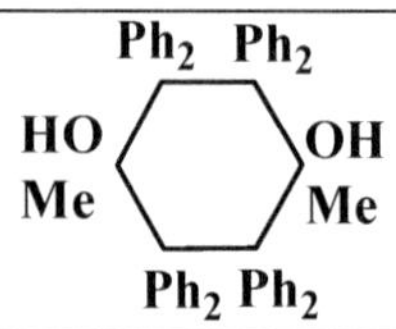	74	–	(mixture) 209-210	–	–	[49,50,52]

Table 1. (Continued)

№	Compound	Yield %	T_{boil}, ^0C (P, mmHg)	T_{melt}, ^0C	$^1H(Me)$ chemical shift		Reference
					cis	*trans*	
17	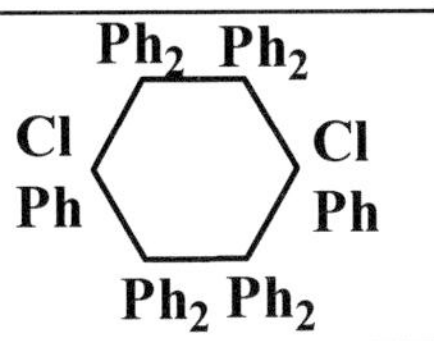	29 (65)	337-340 (0,01)	(mixture) 62-64	–	–	[49,50,52]
18	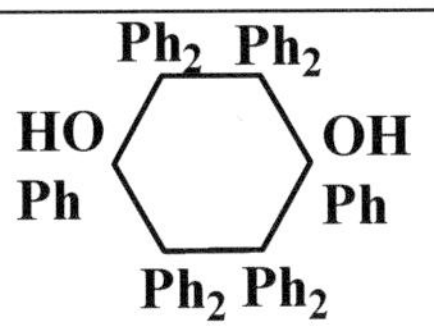	77	–	(mixture) 161-162	–	–	[49,50,52]
19	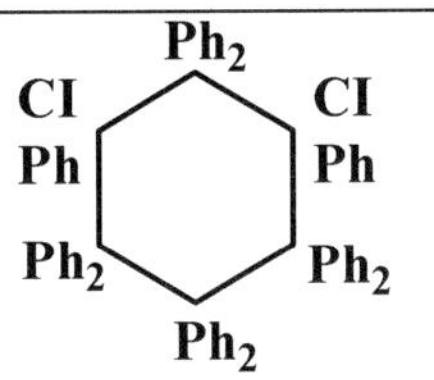	62 (65)	346-352 (0,01)	(mixture) 96-97	–	–	[28,29,50]
20	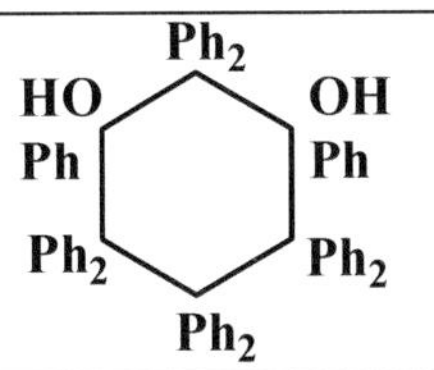	73	–	(mixture) 142-143	–	–	[28,29,50]
21	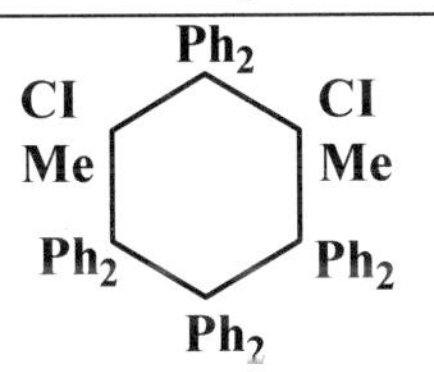	60 (65)	312-315 (0,01)	–	0,34	0,39	[28,29,50]
22	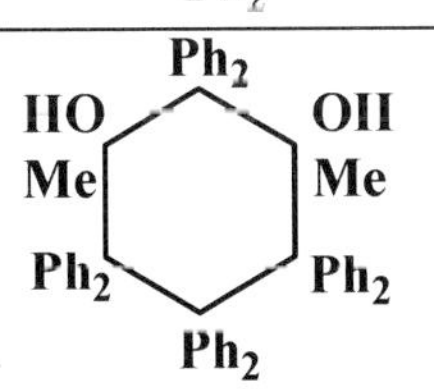	68	–	(mixture) 101-102	–	–	[28,29,50]

Note: * Shown in brackets are yields obtained for the two-stage method.

Hereinafter, cyclotetra-, cyclopenta- and cyclohexasiloxane fragments are denoted as follows:

Cyclotetrasiloxane 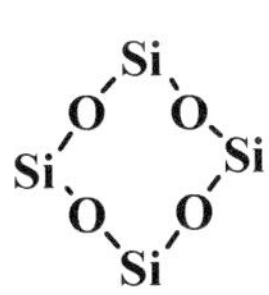

as 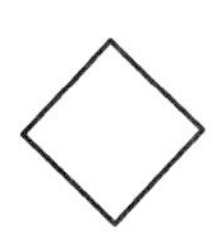

Table 2. Physical and chemical parameters of difunctional methylcyclosiloxanes

№	Compound	Yield, %	T_{boil}, ^0C (P,mmHg)	T_{melt}, ^0C	^1H (CH_3) chemical shift			Reference
					Number of Si atoms	*cis*	*trans*	
1	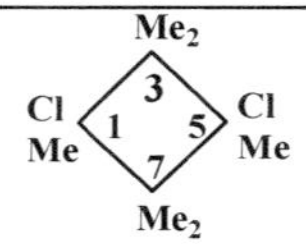	50	201-203 (760)	–	1,5 3,7[a]ь	0,461 0,170 0,296	0,449 0,230	[30]
2	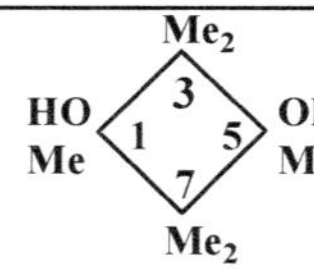	89	115-120 (4-5)	*trans* 119÷121	1,5, 3,7[a]ь	0,065 0,084 0,119	0,067 0,099	[30]
3	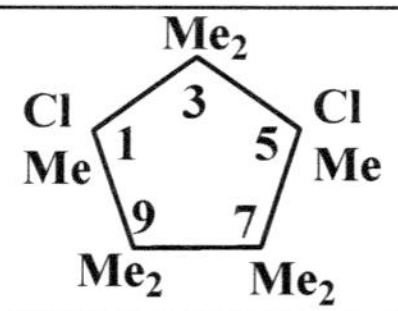	48	222-224 (760)	-	1,5, 3[a] 3б 7,9[a] 7,9б	0,428 0,161 0,279 0,124 0,199	0,433 0,213 0,140 0,175	[30]
4	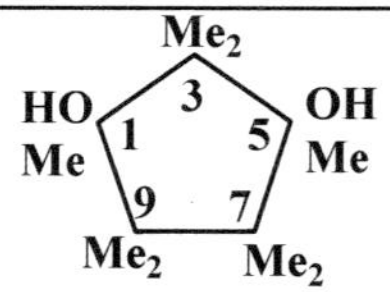	76	89-91 (3-4)	-	1,5 3[a] 3б 7,9[a] 7,9б	0,065 0,091 0,127 0,082 0,110	0,072 0,100 0,100 0,084 0,108	[30]
5	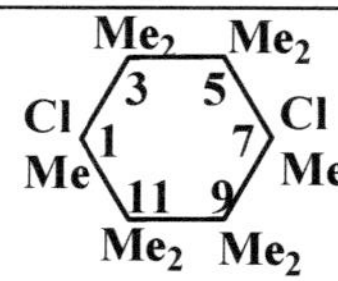	24	55-57 (1-2)	-6÷-8	1,7 3,5[a] 9,11б	0,422 0,129 0,192	0,428 0,140 0,174	[30]
6	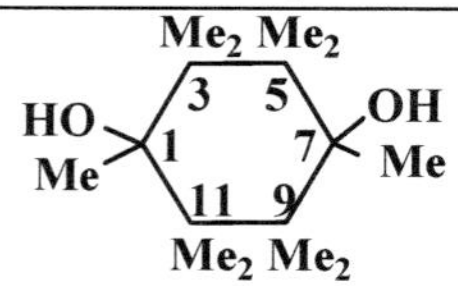	93	–	trans 96÷99 cis 36÷38	1,7 3,5,9,11a 3,5,9,11б	0,065 0,096 0,111	0,083 0,090 0,120	[30]

Table 2. (Continued)

№	Compound	Yield, %	$T_{boil.}$, ^0C (P,mmHg)	$T_{melt.}$, ^0C	^1H (CH_3) chemical shift			Reference
					Number of Si atoms	*cis*	*trans*	
7	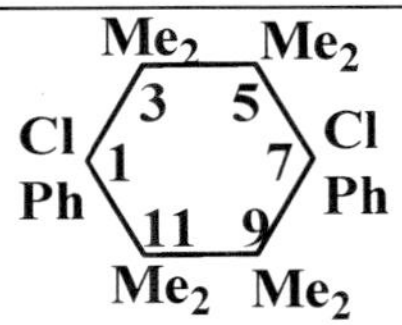	*cis* 13 *trans* 15	175-177 (1-1,5)	– *trans* 90-92	3,5	0,267 0,069	0,192 0,134	[44]
8	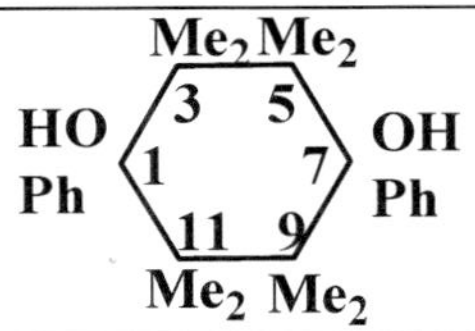	*cis* 89 *trans* 90	- -	*cis* 87-88 *trans* 72-73	3,5	0,078 0,115	0,110 0,130	[88]
9	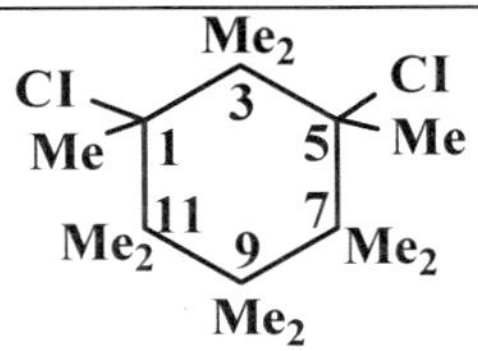	25	82-84 (1-2)	-	3a 3б 1,5 9a 9б 7,11a 7,11б	0,170 0,274 0,426 0,085 0,113 0,123 0,181	0,214 0,436 0,098 0,130 0,170	[30]
10	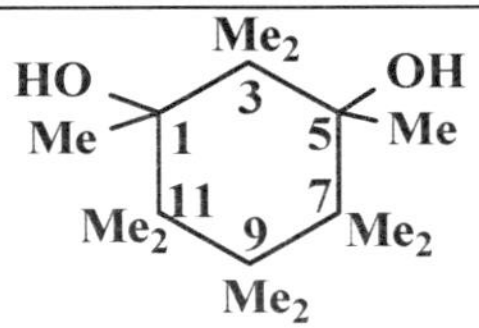	74	102-104 (0,03)	-	3a 3б 1,5, 7,11a 7,11б 9a 9б	0,106 0,136 0,074 0,089 0,125 below the reflex 0,095	 0,110 0,086 0,091 0,119 0,100	[30]

Note: a and b indicate detected signals of nonequivalent methyl groups at silicon atoms.

Table 3. Physical and chemical parameters of difunctional ethylcyclosiloxanes

№	Compound	Yield, %	T_{boil}, ^{0}C (mmHg) T_{melt}, ^{0}C	n_D^{20}	$M^+-C_2H_5$ molecular mass	^{29}Si NMR spectrum (δ, ppm.)			Reference
						Si^1	Si^2	Si^3	
1	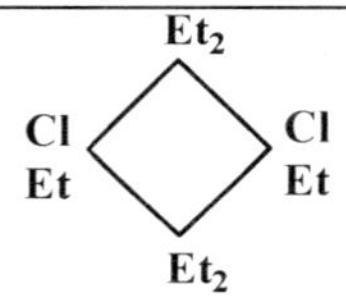	63	113-115 (1-1,5)	1,4355	391 421,6	*cis* -43,14 *trans* -42,87	*cis* -15,45 *trans* -15,03	- -	[31]
2	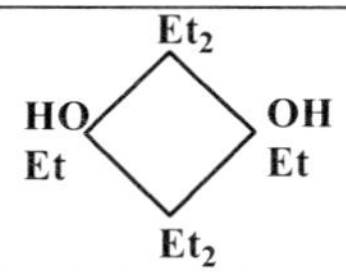	80	32	–	– 384,7	*cis* 54,44 trans -55,53	*cis* -17,70 trans -18,60	- -	[31]
3	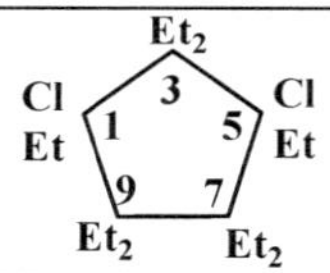	39	128-130 (1-2)	1,4385	493 523,8	*cis* -45,78 *trans* -45,68	*cis* -17,71 *trans* -17,51	*cis* -19,25 *trans* -19,33	[31]
4	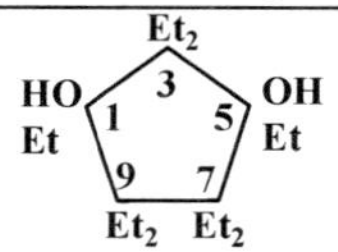	66	120-122 (0,02)	1,4400	– 486,9	*cis* -56,46 *trans* -55,34	*cis* -18,42 *trans* -20,11	*cis* -21,08 *trans* -20,59	[31]
5	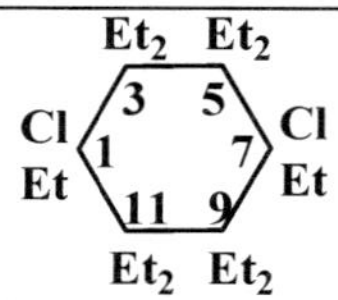	39	170-172 (1-2)	1,4425	595 626	*cis* -46,85 *trans* -46,90	*cis* -19,72 *trans* -19,81	- -	[31]
6	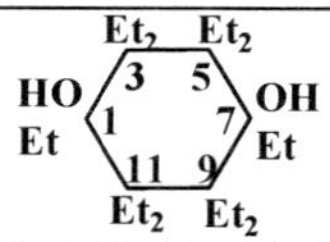	46	160-163 (0,02)	1,4420	– 589,1	-	-	-	[31]

Table 4. Physical and chemical parameters of vinyl-containing organocyclosiloxanes

No.	Structural formula	Yielc %	$T_{boil,}$ ^0C P,(mm Hg) $T_{melt,}$ ^0C	n_D^{20}	d_4^{20}	MR_D^{**}	^1H, ppm. (R) RXSi<O	***References***
1	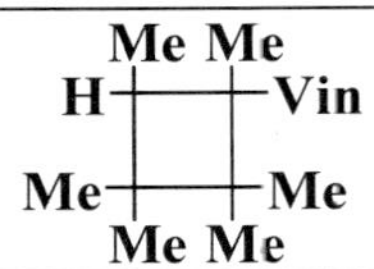	63	58-60 (8)	1,4017	0,9703	74,35 / 32,65	–	[16]
2	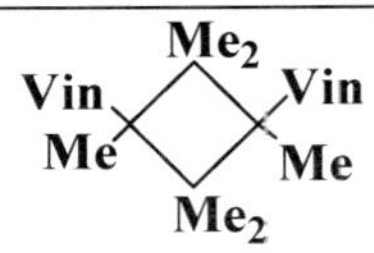	35	70-71 (11)	1,4166	0,9704	83,01 / 82,88	*cis*-0,146 *trans*-0,133 0,106 (Me_2)	[25]
3	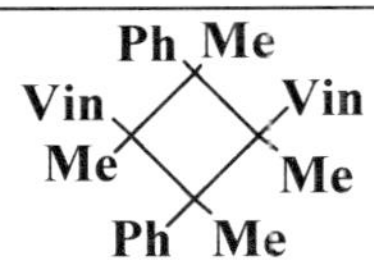	29	150-153 (1-2)	1,5000	1,0779	121,36 / 122,52	–	[25]
4	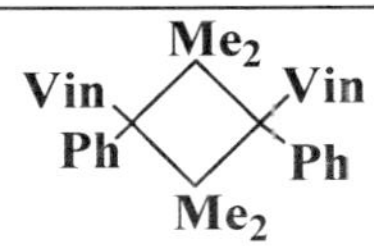	35	137-138 (2)	1,5024	1,0700	122,75 / 122,52	–	[25]
5	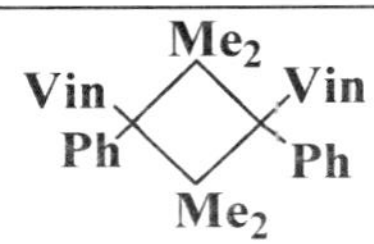	50	242-243 (1-2) 88-89	–	–	–	–	[25]

Table 4. (Continued)

No.	Structural formula	Yield %	T_{boil}, ^0C P,(mm Hg) T_{melt}, ^0C	n_D^{20}	d_4^{20}	MR_D^{**}	^1H, ppm. (R) 	*References*
6	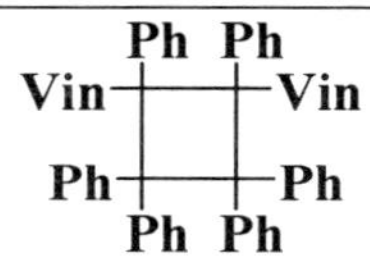	75	255-265 (2) 80-81	–	–	–	–	[26]
7	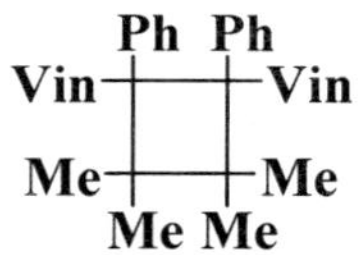	45	130-131 (2)	1,5017	1,0670	<u>122,95</u> 122,52	*cis*-0,15	[26]
8	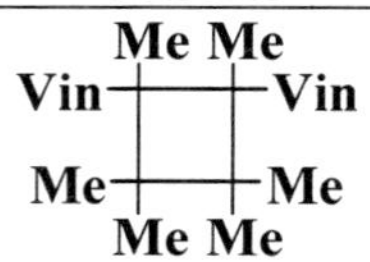	66	87-88 (10)	1,4171	0,9719	<u>82,96</u> 82,88	*cis* -0,156 *trans*-0,141 0,09*(Me_2)	[26]
9	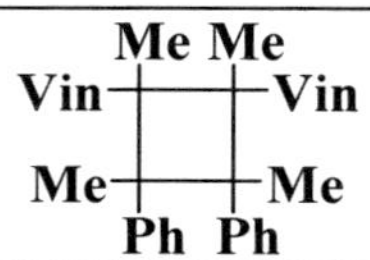	29	150-153 (1-2)	1,5000	1,0970	–	–	[25]
10	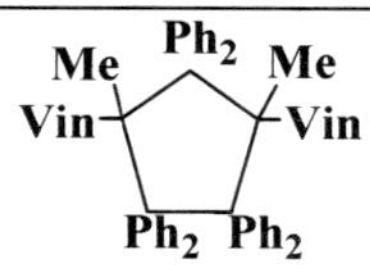	50	255-260 (0,02) 152÷153	–	–	–	*cis* 0,20 *trans* 0,22	[41,53,55]

Table 4. (Continued)

No.	Structural formula	Yield %	$T_{boil,}$ ^0C P,(mm Hg) $T_{melt,}$ ^0C	n_D^{20}	d_4^{20}	MR_D^{**}	^1H, ppm. (R) 	*References*
11	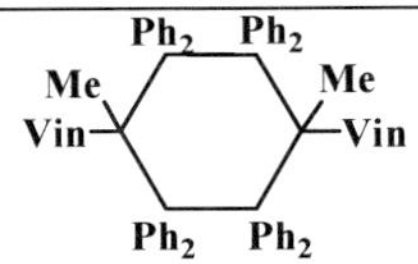	23 (62)	300-305 (0,02) 62-65	–	–	–	*cis* 0,20 *trans* 0,23	[53,56]
12	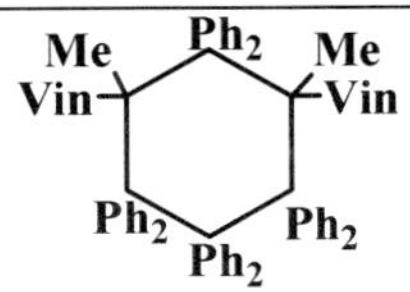	68	290÷294 (0,02) 152÷154	–	–	–	*cis* 0,17 *trans* 0,19	[29,57]
13	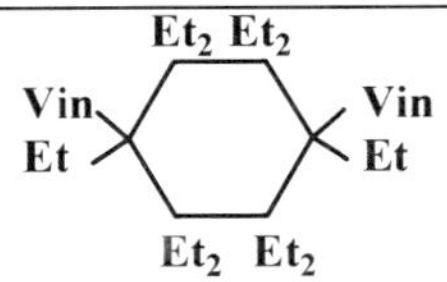	49	180-182 (1)	1,4443	–	–	0,45 q 0,85 t	[47]
14	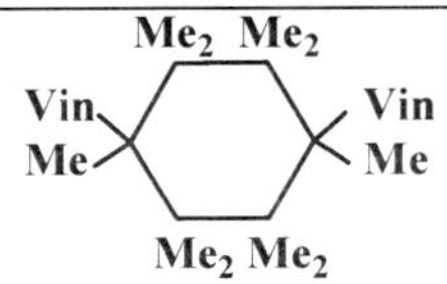	64	83-85 (1-2)	1,4128	–	–	*cis* 0,157 *trans* 0,150	[47]
15	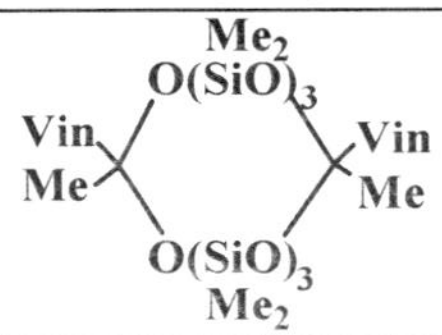	31	119-122 (2)	1,4139	–	–	*cis* 0,23 *trans* 0,25	[47]

Table 5. Physical and chemical parameters of hydride-containing organocyclosiloxanes

	Structural formula	Yield %	T_{boil} ^{0}C P(mm Hg) T_{melt} ^{0}C	n_D^{20}	d_4^{20}	MR_D**	^{1}H, ppm.	***Reference***
1	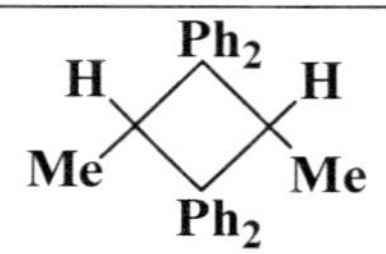	79	245-250 (3-4) 87-88	–	–	–	(mixture)	[25]
2	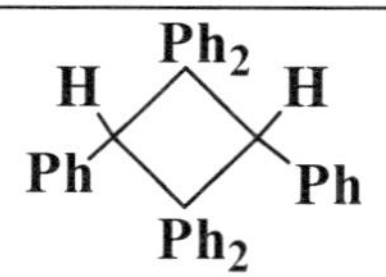	35	270-280 (3) 60-61	–	–	–	(mixture)	[26]
3	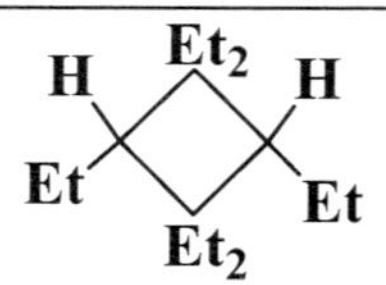	34	95-98 (2)	1,4230	–	–	(mixture) 0,62 к 1,07 т	[47]
4	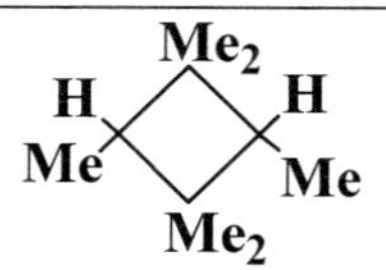	21	57-58 (18) -40	1,3930	0,9680	65,88 65,82	–	[72]
5	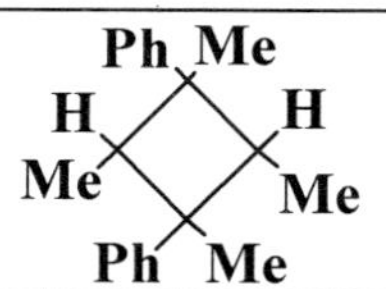	27	140-145	1,4914	1,0907	–	(mixture)	[25]
6	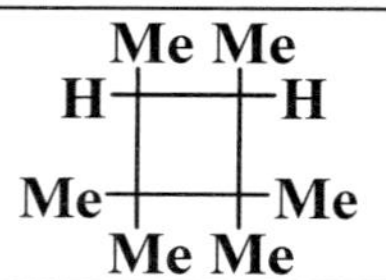	18	154-155	1,3927	0,9709	–	(mixture)	[73]

Table 5. (Continued)

	Structural formula	Yield %	T_{boil} ^0C P(mm Hg) T_{melt} ^0C	n_D^{20}	d_4^{20}	MR_D^{**}	^1H, ppm.	***Reference***
7	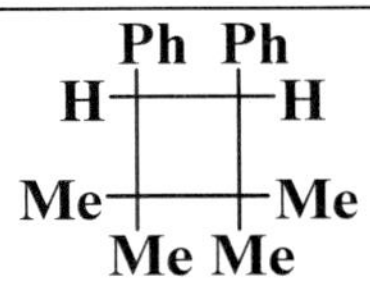	14	137-139 (1-2)	1,4876	1,0698	–	(mixture)	[74]
8	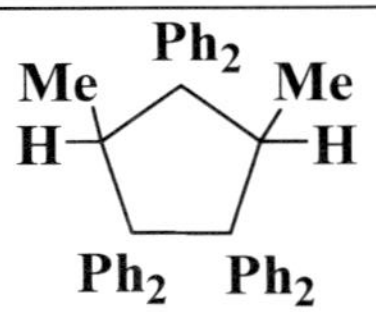	67	235-242 (0,02)	–	–	–	(mixture) *cis*-0,25 *trans*-0,28	[49,52]
9	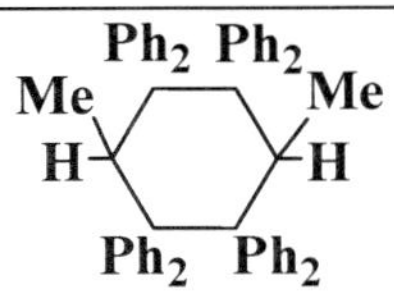	23 (67)	263-267 (0,01) 162-163	–	–	–	(mixture) *cis*-0,26 *trans*-0,30	[47]
10	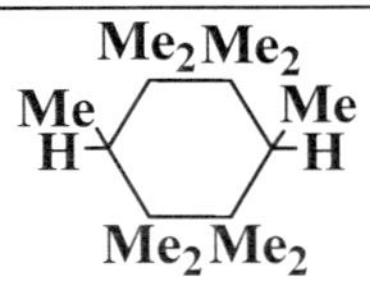	59	66-68 (1)	1,3971	–	–	(mixture) *cis*-0,14 *trans*-0,15	[46]
11	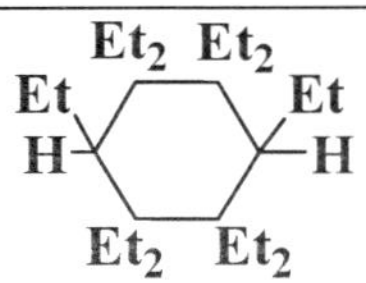	53	175-178 (1)	1,4341	–	–	(mixture) 0,6 q 1,0 t	[46]

Note: * Numerator shows calculated values and denominator – recorded ones.

Table 6. Physical and chemical parameters of tetrafunctional organophenylcyclosiloxanes

№	Compound	Yield, %	$T_{boil.}$, ^0C (P, mm Hg)	$T_{melt.}$, ^0C	Reference
1	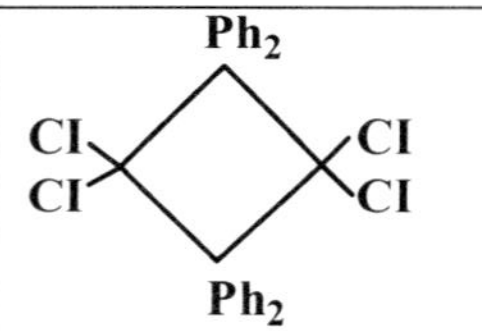	53	206-210 (0,002)	–	[41,58, 75]
2	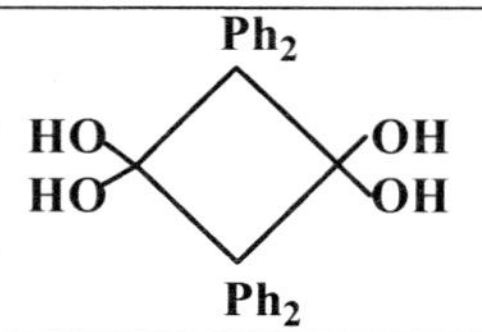	64	–	157-158	[41,58]
3	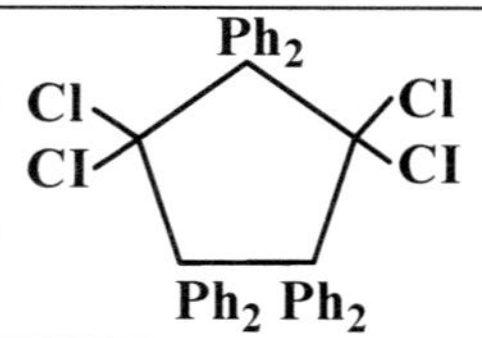	52 (70)	255-260 (0,02)	–	[41,50]
4	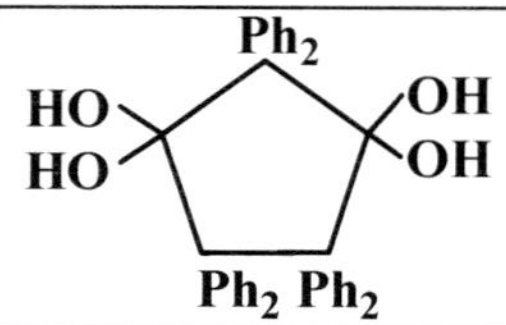	73	–	191-192	[41,50]
5	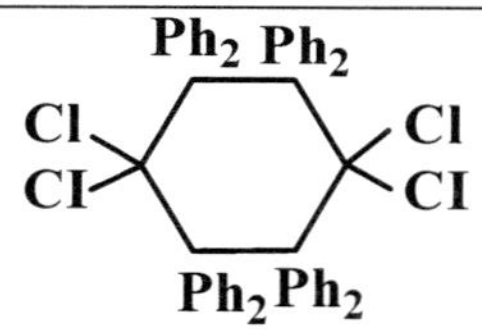	26	325-330 (0,06)	83-87	[49,50,54]

Table 6. (Continued)

№	Compound	Yield, %	$T_{boil.}$, ^0C (P, mm Hg)	$T_{melt.}$, ^0C	Reference
6	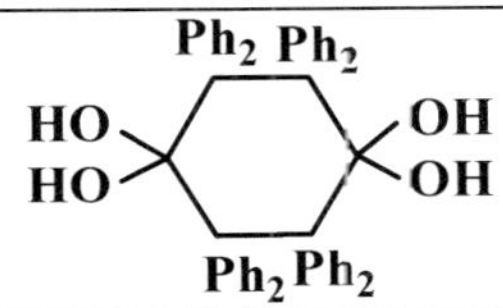	75	–	251-252	[50,54]
7	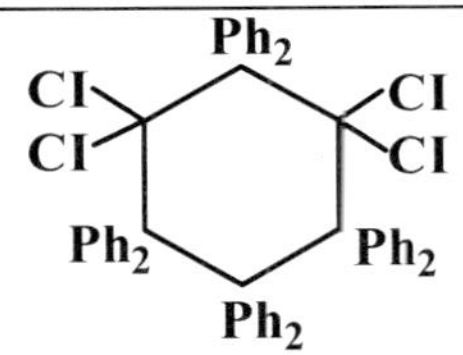	67	307-311 (0,01)	71-73	[29,50]
8	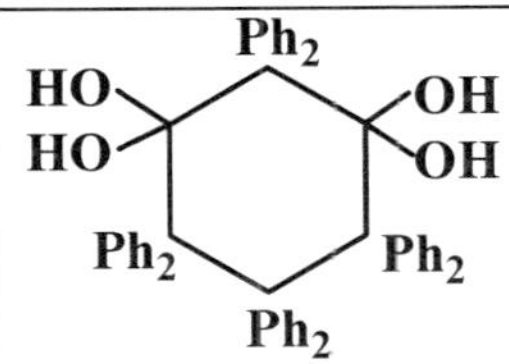	64	–	161-162	[29,50]

Cyclopentasiloxane  as 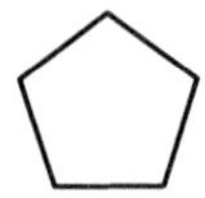

Cyclohexasiloxane 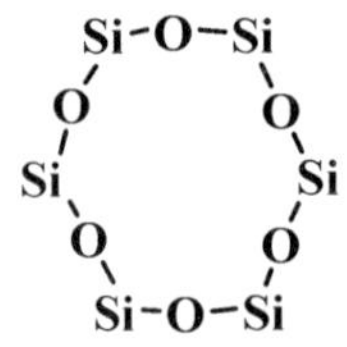 as 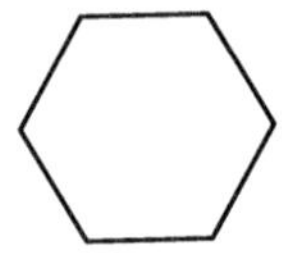

2. Organocyclocarbosiloxanes with Functional Groups at Silicon Atoms

For the purpose of synthesizing organocyclocarbosiloxanes with defined disposition of functional gro-ups at silicon atoms, HFC reaction of 1,2-methylchlorosilyl-substituted ethane with dihydroxydiphe-nylsilane and 1,3-dihydroxytetraphenyldisiloxane in diluted anhydrous toluene solution in the pre-sence of pyridine at 1:1 molar ratio of the initial components was studied [30, 76]. General reaction scheme is as follows:

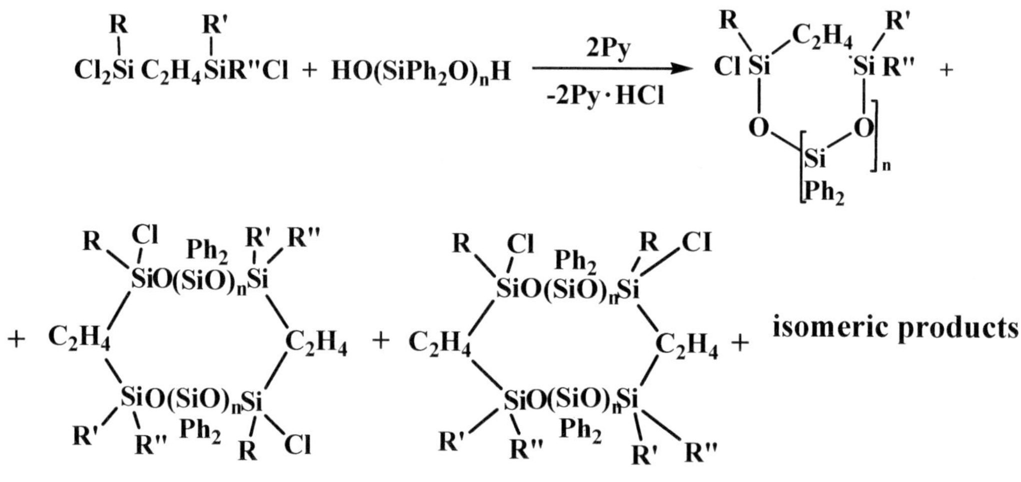

Scheme 32.

Where: n=1, 2; R=R'=Me; R=Cl, R'=R''=Me; R=R'=R'' =Cl.

The yield of organocyclocarbosiloxanes is seriously dependent on the reaction temperature. It is found that if the process proceeds at 20 - 25°C, total yield of cyclocarbosiloxanes does not exceed 70%, the yield of dichlorocarbosiloxanes being equal 55%. Reduction of the reaction temperature to -5 ÷ -10°C allows increase of total yield of cyclocarbosiloxanes up to 85% and dichlorocarbosiloxanes up to 68%. The yield of tetrachlorocyclocarbosiloxane equals 44%. Hydrolysis of di(tetra) chlorocyclo-carbosiloxanes has synthesized corresponded di- and tetrahydroxyderivatives in accordance with the scheme as follows:

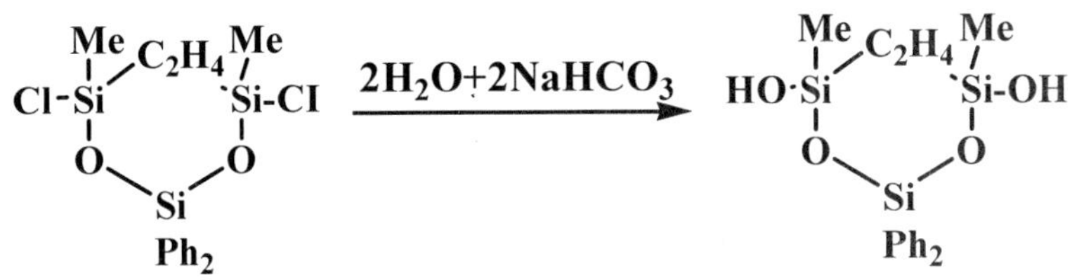

Scheme 33.

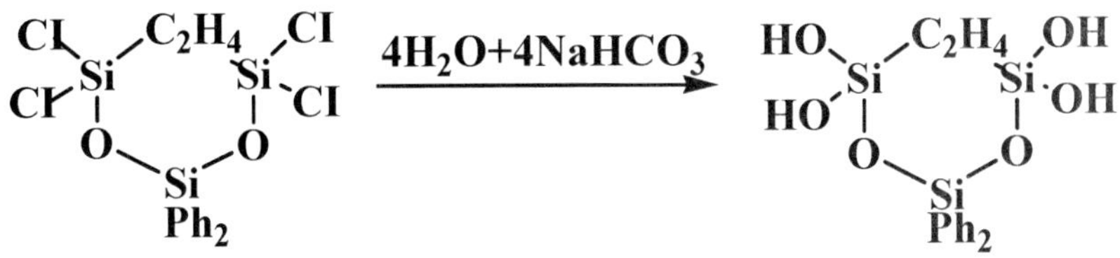

Scheme 34.

HFC reaction of 1,3-bis(dichloromethylsilyl)ethane and 1,3-bis(dichloromethylsilyl) propane with dihydroxydimethylsilane proceeded in diluted anhydrous ether solution, in the presence of pyridine at -5°C. However, the yields of produced methylcyclocarbosiloxanes are low, giving 10 – 22%.

The reaction scheme is as follows [77]:

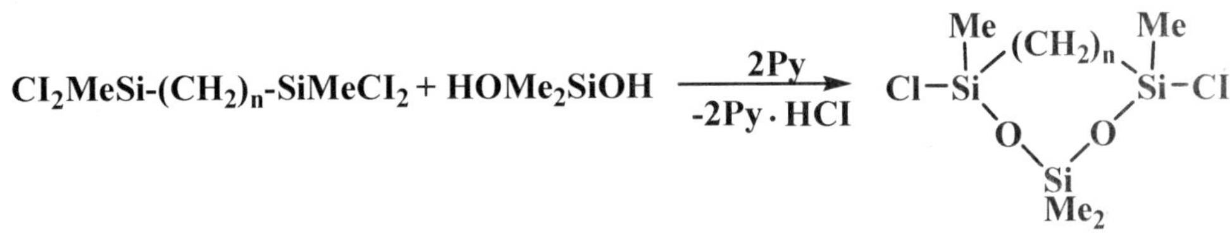

Scheme 35.

Where: $n = 2, 3$.

Extended dichloromethylcyclocarbosiloxane was synthesized by condensation of 1,1,7,7-tetrachloro-1,3,3,5,5,7-carbotetrasilocane with 1,3-dihydroxytetramethyl-disiloxane under similar conditions in accordance with the scheme as follows [77]:

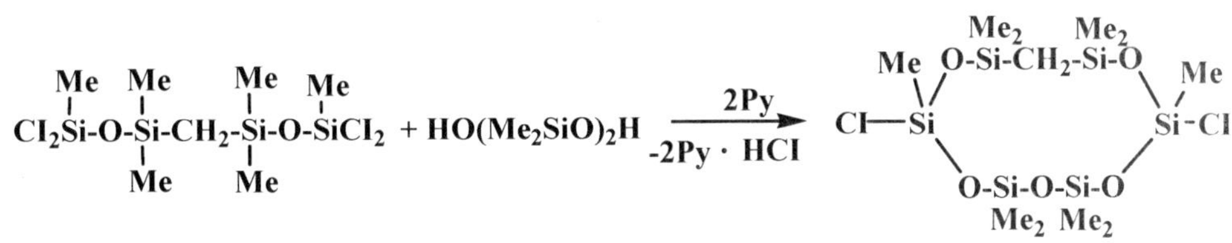

Scheme 36.

Where: R = Me, Ph.

The yield of the above-mentioned dichloromethylcyclocarbosiloxane equals ~40%. Dihydroxy-deri-vatives were synthesized by hydrolysis of dichloromethylcyclo-carbosiloxanes.

Physical and chemical parameters of di- and tetrafunctional organocyclosiloxanes and organocarbo-cyclosiloxanes are shown in Table 7.

Table 7. Physical and chemical parameters of difunctional organocyclocarbosiloxanes

№	Compound	Yield, %	T_{boil}, ^{0}C (mm Hg)	T_{melt}, ^{0}C	^{1}H (CH_3) chemical shift *cis* *trans*	Reference
1	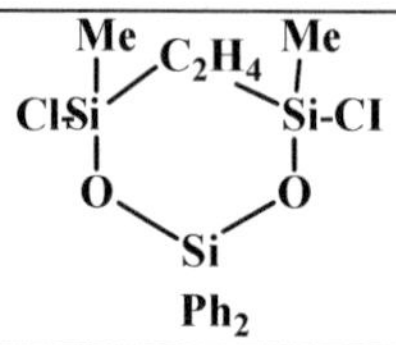	67,9	187-193 (0,01)	77-78	(mixture) *cis*-0,47 *trans*-0,52	[29,47]
2	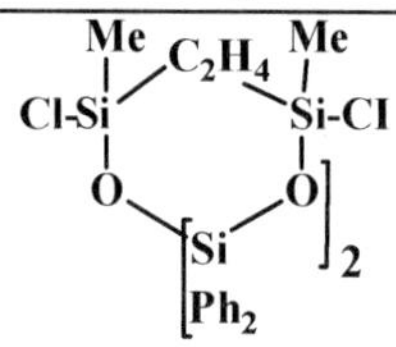	59,4	218-221 (0,02)	110-111	(mixture) *cis*-0,41 *trans*-0,49	[29,47]
3	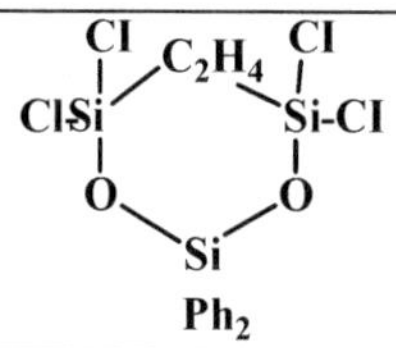	56	150-155 (0,01)	159-161	–	[47]
4	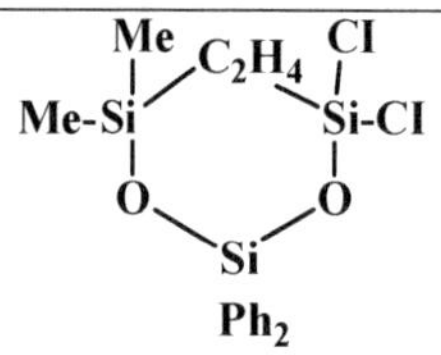	68	149-520 (0,01)	-	0,45 s	[29,47]
5	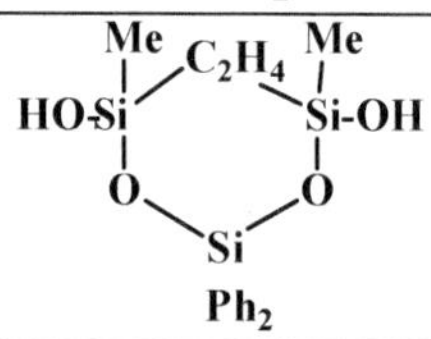	71	-	120-122		[29,47]
6	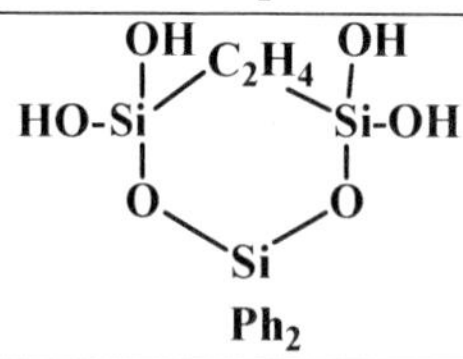	69	–	169-170	–	[29,47]
7	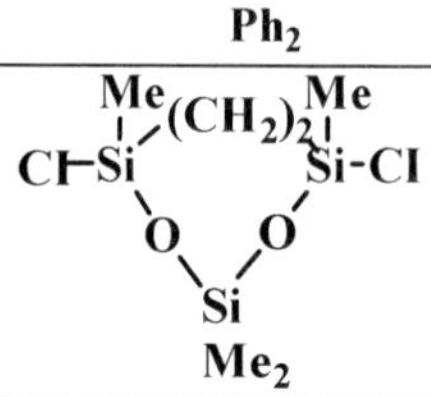	22	61-65 (5)	–	*cis* 0,88 *trans* 1,23	[77]

Table 7. (Continued)

№	Compound	Yield, %	T.boil., ^{0}C (mm Hg)	T.melt., ^{0}C	^{1}H (CH_3) chemical shift *cis* *trans*	Reference
8	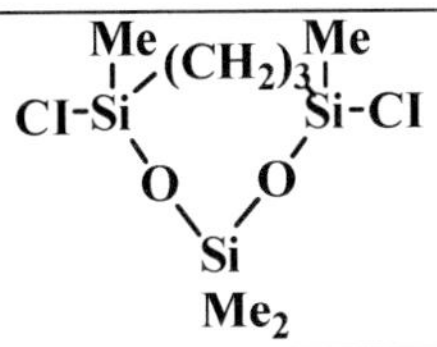	10	66-73 (4-5)	–	*cis* 0,38 *trans* 0,40	[76]
9	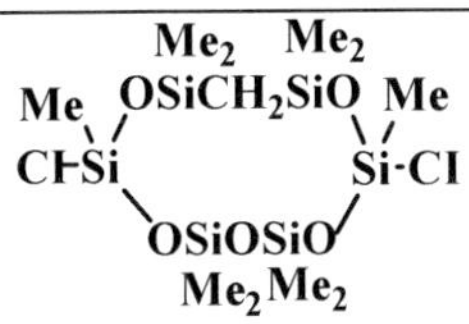	40	99-110	–	MeSiCl* 45,50	[76]
10	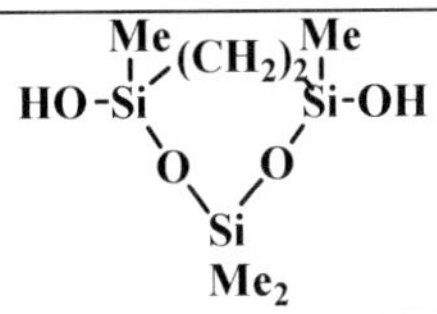	71	80-100 (4-5)		0,05	[76]
11	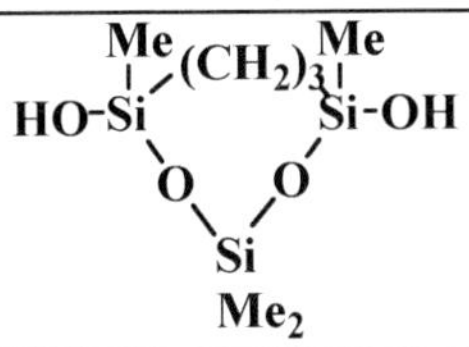	40	130-134		0,04	[76]
12	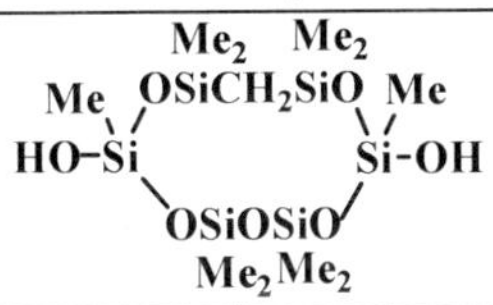	71	–	–	MeSi(OH)* 63,02	[76]

Note: * ^{29}Si spectrum.

Difunctional organosilicon heterocyclic compounds were successfully synthesized by pyrolytic ring formation of symmetrical disiloxanes with phenyl radicals. Pyrolysis in the gas phase in the tem-perature range of 680 - 690°C proceeds with benzene extraction, which leads to synthesis of repre-sentatives of a new class of heterocyclic compounds, which are organochlorine derivatives of 1,3-disila-2-oxaindane. This reaction proceeds by the scheme as follows [78, 79]:

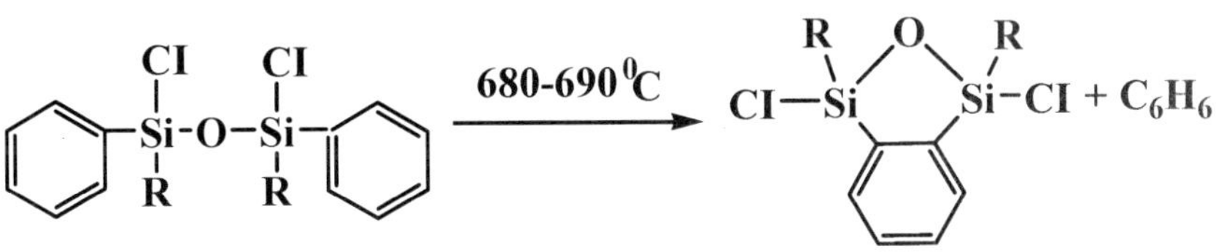

Scheme 37.

Where: $x = 4$.

Studying hydrolysis of disilaoxaindanes, the authors of the work [80] have detected high reactivity of ≡Si-CI bonds, which is explained by decrease of ≡Si-O-Si≡ valence angle in 1,3-disila-2-oxaindane. Lower reactivity of silaindane fragment with phenyl framing at silicon atoms is explained by both steric and inductive effects of the framing groups.

However, hydrolysis of 1,3-dichloro-1,3-disila-1,3-diphenyl-2-oxaindane under soft conditions in the temperature range of -5 ÷ -10°C in the presence of both sodium bicarbonate and aniline has synthe-sized not appropriate dihydroxy-derivatives, but oligomeric α,ω−dihydroxy-1,3-disila-1,3-1,3-diphe-nyloxainadanes. The reaction proceeds in accordance with the scheme as follows [81]:

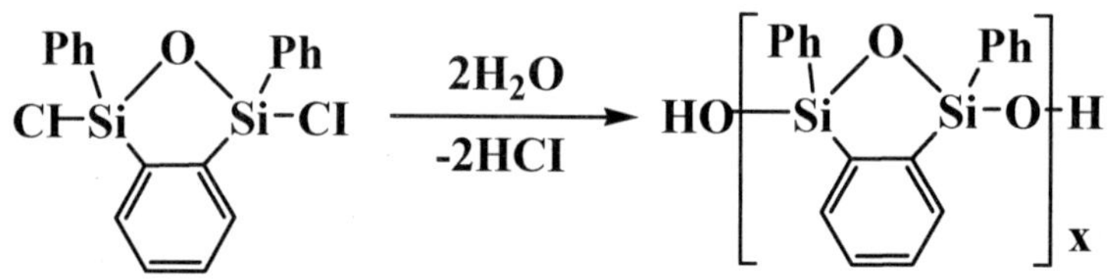

Scheme 38.

Where: x≈4.

Homocondensation of oligomeric dihydroxy-derivatives, boiled in toluene, has synthesized oligomers with higher transformation degree.

It is shown [82] that the gas-phase pyrolysis of *orhto*dichlorosiloxydiphenyl easily synthesizes pen-tatomic silicon-containing heterocycle of 10,10-dichloro-10-sila-9-oxarophenanthrene with the yield of about 60%, which is caused by high reactivity of hydrogen *ortho*-atom.

However, insignificant ring formation with simultaneous synthesis of dibenzofuran did also proceed. The general reaction scheme is as follows:

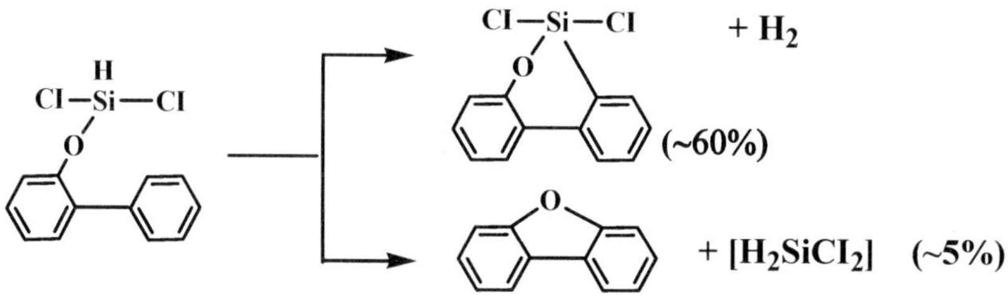

Scheme 39.

Contrary to silaoxaindane, 10,10-dihydroxy-10-sila-oxarophenanthrene and oligomeric dihydroxyde-rivatives have been synthesized in hydrolysis of 10,10-dichloro-10-sila-9-oxarophenanthrene procee-ded in the neutral medium in the presence of aniline. The process scheme is as follows [83]:

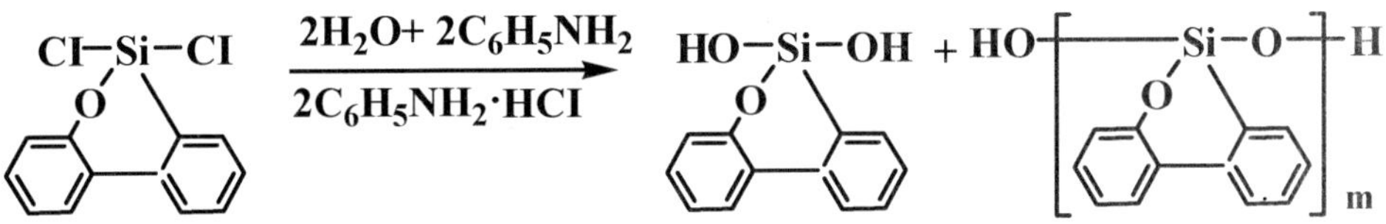

Scheme 40.

Table 8. Physical and chemical parameters of heterocyclic organosiloxanes

№	Compound	Yield, %	T_{melt}, ^0C (P,mm Hg)	d_4^{20}	n_D^{20}	*References*
1	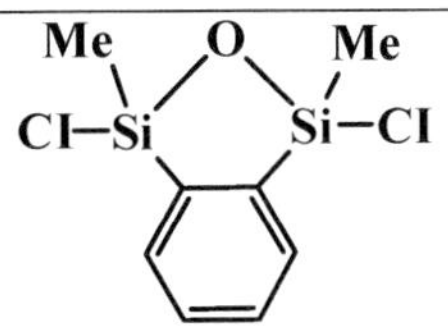	16	86-88 (4)	1,2220	1,5152	[78,79]
2	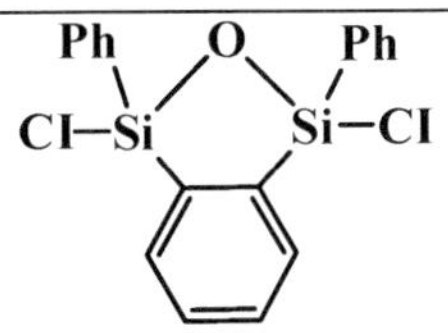	20	220-222 (3)	1,2521	1,6071	[78,79]
3	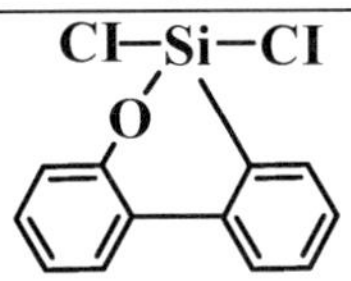	56	160-161 (1) 82-83	-	-	[82]
4	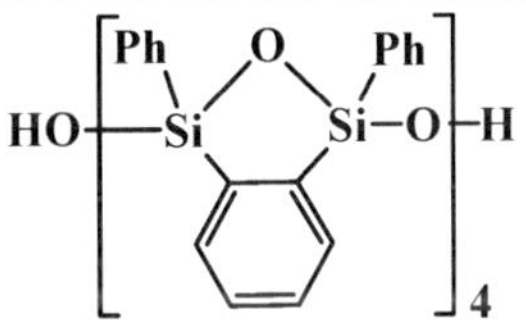	97	48-54	–	–	[80,81]
6	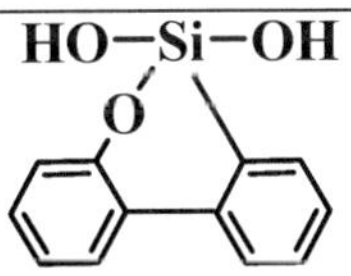	86	236-239	–	–	[83]
7	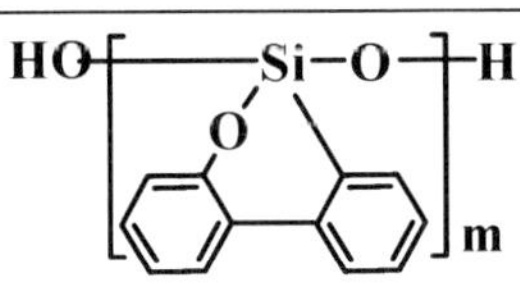	13	120-126	–	–	[83]

It is found that dichlorosilaoxarophenanthrene is not destroyed during boiling with aqueous or alco-holic alkaline, as well as with concentrated hydrochloric acid. At the same time, it is known well that usual Si-O-C bond is easily degraded by alkali and acids, and Si-Car bond is extremely sensitive to the action of acids [84]. Quantitatively without destruction of ≡Si-O- bond, dichlorosilaoxarophe-nanthrene is alkylated and arylated by Grignard reagents, restored by lithium aluminum hydride; when treated by hydrofluoric acid, it is transformed to difluorosilaoxarophenanthrene. Due to the authors' point of view, high chemical stability of silaoxarophenanthrene indicates a definite fragrance of this compound [85]. Some physical and chemical parameters of heterocyclic compounds with functional groups are shown in Table 8.

REFERENCES

[1] Andrianov K.A., Organosilicon Compounds, Moscow, Goskhimizdat, 1955.(Rus)

[2] Andrianov K.A. and Khananashvili L.M., Technology of Organic Monomers and Polymers, Moscow, *Khimia,* 1998. (Rus)

[3] Troenko V.I., Candidate's Dissertation on Chemistry, INEOS AS USSR, Moscow, 1971. (Rus)

[4] Andrianov K.A., Emelianov V.N., and Troenko V.I., Doklady AN SSSR,1969, vol. 189(1), p. 91. (Rus)

[5] Iler R.K., *Ind. Eng. Chem.,* 1947, vol. 39, p. 1384.

[6] Patent No. 2,547,944 (USA), 1951; Iler R.K.

[7] Okawara R., Hotta S., and Watase T., Nippon Kagaky Zasshi, 1954, vol. 75, p. 759.

[8] Takatani T., *J. Chem. Soc. Japan, Pure Chem. Sect.,* 1953, vol. 74, p. 890.

[9] Takatani T., J. *Chem. Soc. Japan, Pure Chem. Sect.,* 1953, vol. 73, p. 409.

[10] Sprung M.M. and Guenter F.O., *J. Amer. Chem. Soc.,* 1955, vol. 77, p.4173.

[11] Okawara R., Minami G., and Oku Z., *Bull. Chem. Soc.* Japan, 1958, vol. 31, p. 22.

[12] Okawara R., Hotta S., and Watase T., *J. Chem. Soc.* Japan, 1953, vol. 27, p.785.

[13] Andrianov K.A. and Volkova L.M., *Vysokomol. Soedin.,* 1961, vol. 3A(10), p. 1580. (Rus)

[14] Andrianov K.A., Fromberg M.B., and Arsenieva E.N., *Zh. Prikl. Khim., 1970,* vol. 43(9), p. 2053. (Rus)

[15] Zhdanov A.A., Andrianov K.A., and Malyikhin A.P., Doklady AN SSSR, 1973, vol. 211(5), p. 1104. (Rus)

[16] Andrianov K.A., Sidorov V.I., Zaitseva M.G., and Khananashvili L.M., *Khimia Geterotsikl. Soe-din.,* 1967, No. 1, p. 32. (Rus)

[17] Brown J.F., *J. Amer. Chem. Soc.,* 1965, vol. 87, p. 4317.

[18] Klementiev I.Yu., Shklover V.E., Kulish M.A., Tikhonov V.S., and Volkova E.V., Doklady AN SSSR, 1981, vol. 259(6), p. 1371. (Rus)

[19] Andrianov K.A., Nogaideli A.I., Tskhovrebashvili V.S., Nakaidze L.I., and Zhinkin D.Ya., Doklady AN SSSR, 1975, vol. 223(1), p. 603. (Rus)

[20] Andrianov K.A., Pavlova S.A., Tverdokhlebova I.I., Emelianov V.N., Larina T.A., and Rabkina A.Yu., *Vysokomol. Soedin.,* 1972, vol. 14(10), p. 1580. (Rus)

[21] Andrianov K.A., Blokhina O.G., and Volkova L.M., *Vysokomol. Soedin.,*1976, vol. 46(7), p. 1537. (Rus)

[22] Andrianov K.A., Volkova L.M., and Tartakovskaya L.M., Izv. AN SSSR, *Ser. Khim.,* 1963, No. 1, p. 294. (Rus)

[23] Andrianov K.A. and Volkova L.M., Izv. AN SSSR, *Ser. Khim.,* 1962, No. 1,p. 264. (Rus)

[24] Patent No. 3,340,289 (USA), 1967.

[25] Zhdanov A.A., Andrianov K.A., Astapova T.A., and Vinogradova G.F., Izv. AN SSSR, *Ser. Khim.,* 1973, No. 11, p. 2628. (Rus)

[26] Zhdanov A.A., Andrianov K.A., Malyikhin A.P., Dubchak I.L., and Strelkova T.V., Izv. AN SSSR, *Ser. Khim.,* 1976, No. 5, p. 1113. (Rus)

[27] Andrianov K.A., Zachernyuk A.B., and Danielov S.A., *Khimia Geterotsikl.Soed.,* 1972, No. 7, p. 893. (Rus)

[28] Achelashvili V.A., Mukbaniani O.V., Khananashvili L.M., and Kikolaldze V.S., *Zh. Obshch. Khim., 1*986, vol. 56(7), p. 1530. (Rus)

[29] Achelashvili V.A., Candidate's Dissertation on Chemistry, Iv. Javakhishvili Tbilisi State Uni-versity, 1982, Tbilisi, Georgia. (Rus)

[30] Makarova N.N., Petrova I.M., Godovski Yu.K., Lavrukhin B.D., and Zhdanov A.A., Doklady AN SSSR, 1983, vol. 269(6), pp. 1368 – 1372. (Rus)

[31] Makarova N.N., Lavrukhin B.D., Turkeltaub G.N., Kuzmin N.N., Petrova I.M., and Matukhina E.V., Izv. AN SSSR, *Ser. Khim.,* 1989, No. 6, p. 1351.(Rus)

[32] Hyde J.F., *J. Amer. Chem. Soc.,* 1953, vol. 75, p. 2167.

[33] Sobolevski M.V., Chistyakova L.A., Nazarova D.V., and Kurilina, *Plast. Massy,* 1962, No. 10, p. 17. (Rus)

[34] Suvorov A.L. and Spasski S.S., Trudy Inst. Khimii AN SSSR, Urals filial, 1966, No. 13, p. 25. (Rus)

[35] Patent No. 1,329,180 (UK), 1974.

[36] Patent No. 1,246,945 (France), 1963; C.A., 1964, vol. 60, 12052.

[37] Mukbaniani O.V., Candidate's Dissertation on Chemistry, Iv. Javakhishvili Tbilisi State Uni-versity, 1977, Tbilisi, Georgia. (Rus)

[38] Andrianov K.A., Emelianov V.N., Muzafarov A.M., and Tot A., Doklady AN SSSR, 1975, vol. 222(3), p. 603. (Rus)

[39] Koiava N.A., Mukbaniani O.V., Khananashvili L.M., and Kutateladze G.Sh., Soobshch., AN GSSR, 1980, vol. 97(3), p. 617. (Rus)

[40] Koiava N.A., Mukbaniani O.V., Khananashvili L.M., and Tsitsishvili V.G., *Zh. Obshch. Khim.,* 1980, vol. 50(8), p. 1793. (Rus)

[41] Koyava N.A., Candidate's Dissertation on Chemistry, Iv. Javakhishvili Tbilisi State University, 1982, Tbilisi, Georgia. (Rus)

[42] Lebedev E.P. and Fedorov A.D., *Zh. Obshch. Khim.,* 1978, vol. 48, p. 931.(Rus)

[43] Makarova N.N., Petrova I.M., Godovski Yu.K., Lavrukhin B.D., and Zhdanov A.A., Doklady AN SSSR, 1983, vol. 269(6), p. 1368. (Rus)

[44] Makarova N.N.Lavrukhin B.D., Timofeeva T.V., and Zelencheva V.N., Izv. AN SSSR, *Ser. Khim.,* 1985, No. 5, p. 1114. (Rus)

[45] Furmanova H.G., Andrianov K.A., and Makarova N.N., *Zh. Strukt. Khim.,* 1987, vol. 28, p. 113. (Rus)

[46] Lavrukhin B.D., Makarova N.N., and Zhdanov A.A., Izv. AN SSSR, *Ser. Khim.,* 1986, No. 3, p. 559. (Rus)

[47] Makarova N.N., Astapova T.V., and Lavrukhin B.D., Izv. RAN, *Ser. Khim.,*1996, No. 4, p. 958. (Rus)

[48] Makarova N.N., Petrova I.M., and Zhdanov A.A., Doklady AN SSSR, 1982, vol. 262(3), p. 619. (Rus)

[49] Meladze S.M., Mukbaniani O.V., and Khananashvili L.M., *Zh. Obshch. Khim.,*1980, vol. 51(7), p. 1624. (Rus)

[50] Mukbaniani O.V., Doctor's Dissertation on Chemistry, Iv. Javakhishvili Tbilisi State University, 1993, Tbilisi, Georgia. (Rus)

[51] Yuzhelevski Yu.A., Kurlova T.V., Kogan B.G., and Suvorova M.V., *Zh. Obshch. Khim.,* 1972, vol. 42(9), p. 2006. (Rus)

[52] Koiava N.A., Mukbaniani O.V., and Khananashvili L.M., Thes. Rep. V All-Union Conference on Chemistry and Application of Organosilicon Compounds, Tbilisi, 1980, part 1, p. 20. (Rus)

[53] Koiava N.A., Mukbaniani O.V., Khananashvili L.M., and Tsitsishvili V.G., *Zh. Obshch. Khim.,* 1981, vol. 51(1), p. 130. (Rus)

[54] Meladze S.M., Candidate's Dissertation on Chemistry, Iv. Javakhishvili Tbilisi State University, 1980, Tbilisi, Georgia. (Rus)

[55] Karchkhadze M.G., Mukbaniani N.O., Samsonia A.Sh., Tkeshelashvili R.Sh., Kvelashvili N.G., Chogovadze T.V., and Khananasvili L.M., Bulletin of Georgian Academy of Sciences, 1998, vol. 158(1), p. 75. (Rus)

[56] Karchkhadze M.G., Mukbaniani N.O., Khananasvili L.M., Meladze S.M., Kvelashvili N.G., and Doksopulo T.P., Intern. *J. Polym. Mater.,* 1998, vol. 41, p. 89.

[57] Mukbaniani N.O., Karchkhadze M.G., Samsonia A.Sh., Tkeshelashvili R.Sh., and Khananashvili L.M., Bulletin of Georgian Academy of Science, 1999, vol. 160(1), p. 81. (Rus)

[58] Andrianov K.A., Zhdanov A.A., Kurasheva N.A., and Khyinku E.S., Doklady AN SSSR, 1977, vol. 236, p. 354. (Rus)

[59] Andrianov K.A., Zhdanov A.A., Kurasheva N.A., and Khyinku E.S., Doklady AN SSSR, 1978, vol. 240(3), p. 591. (Rus)

[60] Andrianov K.A., Zhdanov A.A., Kurasheva N.A., and Khyinku E.S., Thes. Rep. V Intern. Symp. Chemistry of Organosilicons, FRG, Karlsrue, 1978, p. 72.

[61] Kurasheva N.A. and Khyinku E.S., Thes. Rep. V All-Union Conference on Chemistry and Application of Organosilicon Compounds, Tbilisi, 1980, p.155. (Rus)

[62] Zhdanov A.A., Kurasheva N.A., and Khyinku E.S., *Izv. AN KazSSR,* 1981, No. 1, p. 38. (Rus)

[63] Andrianov K.A. and Zachernyuk A.B., *Khim. Geterotsikl. Soed.,* 1972, No.8, p. 1183. (Rus)

[64] Kurasheva N.A., and Khyinku E.S. Thes. Rep. V All-Union Conference on Chemistry and Appli-cation of Organosilicon Compounds, Tbilisi, 1980, part 1, p.155. (Rus)

[65] Andrianov K.A., Nogaideli A.I., Slonimski G.L., Levin Yu.V., Makarova N.N., Kvachev Yu.P., and Mukbaniani O.V., *Vysokomol. Soedin.,* 1976, vol.18B, p. 359. (Rus)

[66] Zachernyuk A.B., Burlova E.A., and Zhdanov A.A., *Zh. Obshch. Khim.,*1985, vol. 55(6), p. 1368. (Rus)

[67] Zachernyuk A.B., Zhdanov A.A., and Shkvoler V.E., *Zh. Obshch. Khim.,* 1987, vol. 57(3), p. 562. (Rus)

[68] Zachernyuk A.B., Candidate's Dissertation on Chemistry, INEOS, AS USSR, Moscow, 1974. (Rus)

[69] Andrianov K.A., Zachernyuk A.B., and Danielov S.A., *Khim. Geterotsikl. Soed.,* 1972, No. 7, p. 893. (Rus)

[70] Andrianov K.A. and Zachernyuk A.B., *Vysokomol. Soedin.,* 1974, vol.16(6), p. 1436. (Rus)

[71] Burlova E.A., Zachernyuk A.B., Vinogradov G.V., and Yanovski Yu.G., Thes. Rep. V All-Union Conference on Chemistry and Application of Organosilicon Compounds, Tbilisi, 1980, part 2, p. 575. (Rus)
[72] Sokolov N.N., *Zh. Obshch. Khim.*, 1959, vol. 29(1), p. 258. (Rus)
[73] Okawara R., Takahashi U., and Sakiiyama M., *Bull. Chem. Soc. Japan,* 1957, vol. 30, p. 608.
[74] Andrianov K.A., Nogaideli A.I., Khananashvili L.M., and Nakaidze L.I. Izv., AN SSSR, *Ser. Khim.,* 1968, vol. 4, 828. (Rus)
[75] Andrianov K.A., Emelianov V.N., Muzafarov A.M., and Tot A., Doklady AN SSSR, 1976, vol. 226, p. 827. (Rus)
[76] Inaridze I.A., Candidate's Dissertation on Chemistry, Iv. Javakhishvili Tbilisi State University, 1994, Tbilisi, Georgia. (Rus)
[77] Chizhova N.V., Makarova N.N., Godovski Yu.K., and Buzin A.I.,*Vysokomol. Soedin.,* 2000, vol. 42A(11), p. 1797. (Rus)
[78] A.C. No. 410,018 (USSR), 1974; Chernyishev E.A., Komalenkova N.G.,Klochkova T.A., Kuzmina G.M., and Bochkarev V.N., *Otkr., izobr.,prom., obr., tov. znaki,* 1974, No. 1. (Rus)
[79] Chernyishev E.A., Komalenkova N.G., Klochkova T.A., Kuzmina G.M., and Bochkarev V.N., *Zh. Obshch. Khim.,* 1975, vol. 45(10), p. 2299. (Rus)
[80] Chernyishev E.A., Belkina T.V., Nikitin V.S., Komalenkova N.G.,Shapatin A.S., and Zhinkin D.Ya., *Zh. Obshch. Khim.,* 1978, vol. 48(3), p.630. (Rus)
[81] Mukbaniani O., Matsaberidze M., Karchkhadze M., and Khananashvili L., Russian Polymer News, 2002, vol. 7(1), p. 7.
[82] Chernyishev E.A., Krasnova T.L., and Shchepinov S.A., *Zh. Obshch.Khim.,* 1970, vol. 40(8), p. 1735. (Rus)
[83] Samsonia A.Sh., Mukbaniani O.V., and Karchkhadze M.G., Georgian Engineering News, 2000, No. 2, p. 142.
[84] Dewar M.J.S. and Dietz R., *J. Chem. Soc.,* 1960, p. 1344.
[85] Dewar M.J.S., Cubba V.P., and Pettit R., *J. Chem. Soc., 1958*, p. 3073.

In: New Topics in Monomer and Polymer Research
Editors: A. D'Amore and G. Zaikov, pp. 113-134
ISBN: 1-60021-436-3

Chapter 8

HETEROBI(TRI, TETRA)CYCLIC ORGANOSILICON COMPOUNDS WITH FUNCTIONAL GROUPS

O. Mukbaniani*[*1]*, G. Zaikov*[2]*,
T. Tatrishvili*[1]* and N. Mukbaniani
[1] I.Javakhishvili Tbilisi State University,
I. Chavchavadze Ave.,1, Tbilisi; 380028, Georgia
[2] Institute of Chemical Physics, Russian Academy of Sciences,
Kosygin Str.4, 1117334 Moscow, Russia

INTRODUCTION

There is some information in the literature about synthesis of functional group containing organocyc-losiloxane. Synthesis of organocyclosiloxane with different disposition of functional groups at atoms of silicon is mainly reduced to two methods: hydrolytic condensation and co-condensation of chlorine-containing organosilanes and alkoxyorganosilanes and hetero functional condensation (HFC) reaction of silanes with siloxanes containing functional chlorine-, alkoxy-, amino and hydroxyl groups or to modification of previously prepared organocyclosiloxanes. The first series of works has used co-hydro-lysis products of dimethyldichlorosilane with phenyltrichlorosilane [1,2], dimethyldichlorosilane with methyltrichlorosilane [3] or methylphenyldichlorosilane with phenyltrichlorosilane [4]. During these reactions, organocyclosiloxanes with low yields and statistical arrangement of functional groups at silicon are obtained. From these two methods of synthesis of functional group containing organocyc-losiloxanes, most of all acceptable reaction is HFC.

[*] Correspondence should be addressed: mukbaniani@ictsu.tsu.edu.ge

1. Synthesis of Heterocyclic Organobi-, Organotri- and Organotetracyclosiloxanes and with Functional Groups

HFC reactions of *cis*-1,3,5,7-tetrahydroxy-1,3,5,7-tetraphenylcyclotetrasilocane (tetrole, T-1) and 1,1,5,5-tetrahydroxy-3,3,7,7-tetraphenylcyclotetrasiloxane (T-2) with various organochlorosiloxane and siloxanes for the purpose of obtaining polycyclic compounds with functional groups and some their transformations are discussed in this Chapter.

When polycondensation of tetrole with organodichlorosilanes is performed with 1:4.5 and 1:5.0 ratios of initial components in the presence of hydrogen chloride acceptor, spatial hindrances for interchain reaction due to *cis*-conformation of reactive hydroxyl groups may be expected. At the same time, in the interaction of tetrole with organochlorosilanes at 1:4.0 ratios of initial components or higher, the reaction may possibly proceed in two directions: formation of polycyclic compounds by intramolecular cyclization and synthesis of more complex structures at intermolecular condensation of intermediate products.

At the first stage model condensation reactions of tetrole with trimethylchlorosilane were studied [5]. Under the following reaction conditions: 1:5 ratio of initial components, temperature range 20+25°C (under soft conditions), dilution C = 0.03 g/ml in ether solution, the presence of acceptor –pyridine, attempts to substitute all four hydroxyl groups to trimethylsiloxy-groups have failed. As a result, 1-hy-droxy-3,5,7-tris(trimethylsiloxy)-tetraphenylcyclotetrasiloxane (71%) and reaction products of intermo-lecular condensation of high transformation degree were obtained. The reaction proceeded as follows:

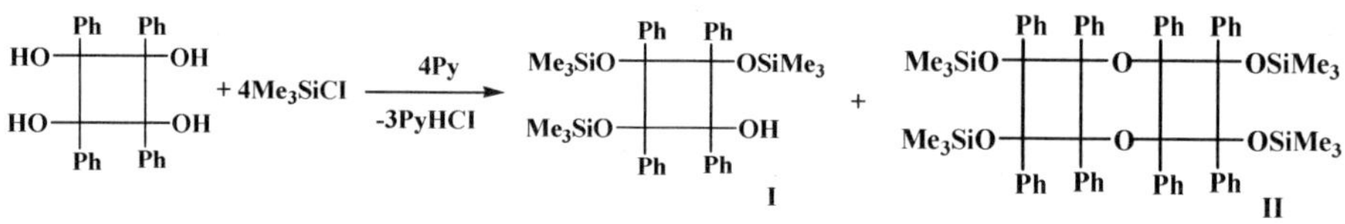

Scheme 1.

Thus, in spite of condensation proceeding with chlorosilane excess, attempts to substitute all four hydroxyl groups by trimethylsiloxy-groups failed under the above-mentioned conditions due to steric hindrances, and besides HFC product, homofunctional condensation product of the structure II is obtained.

^{1}H NMR spectrum of 1-hydroxy-3,5,7-tris(trimethylsiloxy)tetraphenylcyclotetrasiloxane displays a singlet reflex with chemical shift at δ=0.20 ppm. ^{1}H NMR spectrum of the product of the structure II displays expanded reflex for methyl protons, which is explained by formation of tricyclic compounds with different spatial structures:

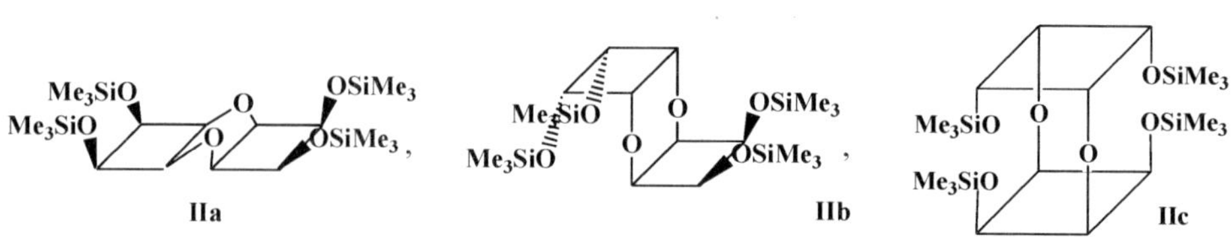

Compounds of the structure II[1] can be synthesized by intermolecular, homocondensation of products containing hydroxyl groups only. As a consequence, the authors suggest that mono- and dihydroxy-derivatives are formed as intermediate compounds:

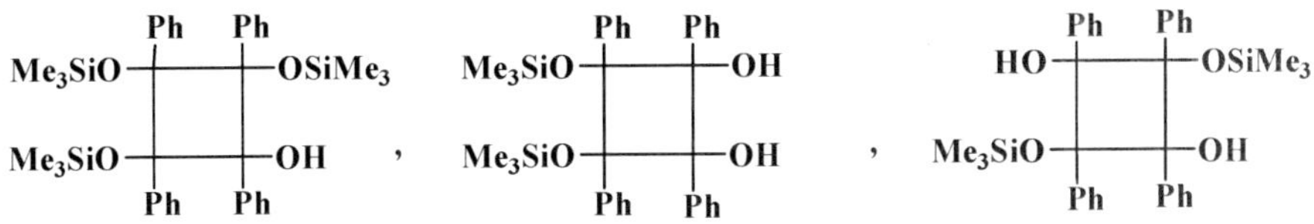

a b c

with future condensation of tetrole dihydroxy-derivatives, which leads to formation of tricycles of the structure II (a – c).

As the temperature mode is changed, i.e. when at the final stage the reaction proceeds at 80°C in benzene (under rigid conditions), it proceeds completely with complete substitution of all hydroxyl groups by trimethylsiloxy-groups. As a result, 1,3,5,7-tetrakis(trimethylsiloxy)-1,3,5,7-tetraphenyl-cyclotetrasiloxane with 78% yield was formed:

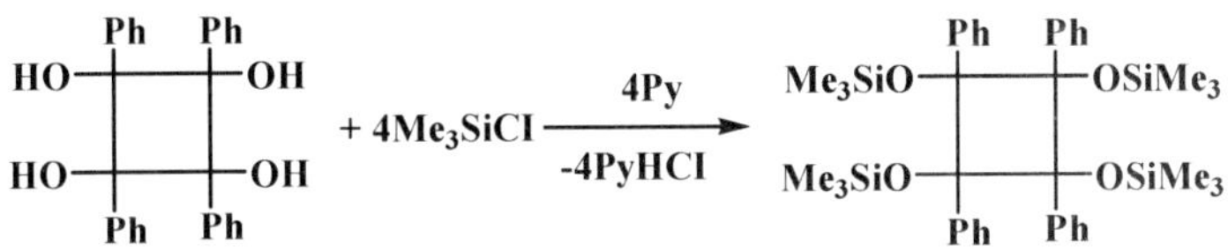

III

Scheme 2.

It should be noted that intermolecular condensation products were not produced in this case. The compound of the structure III was produced in HFC reaction of the compound of structure I with trimethylchlorosilane in benzene at 1:1.5 ratios of initial components. ^{1}H NMR spectrum of the compound of the structure III displays a singlet with chemical shift δ=0.19 ppm for methyl protons due to their equivalence.

By analogy, HFC reaction of tetrole with dimethylchlorosilane proceeds under rigid conditions and at 1:4 ratios of initial components for soft conditions. In this case, three hydroxyl groups only are sub-stituted, and the fourth one is preserved. However, contrary to reaction (1), intermolecular conden-sation product cannot be produced by fractionation of the reaction products. As a consequence, the reaction proceeds in accordance with the following scheme:

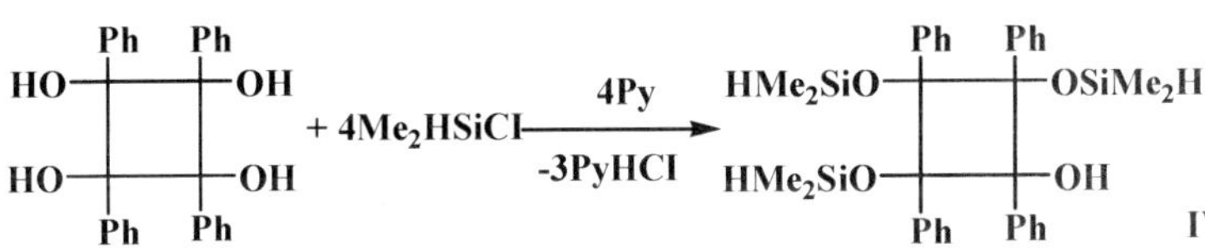

IV

Scheme 3.

---

[1] Hereinafter cyclotetrasiloxane fragments 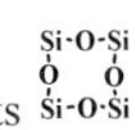 are denoted as: 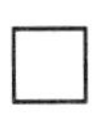

Further HFC reaction of compound 1 (Table 1) of structure IV with dimethylchlorosilane gives an opportunity to substitute the fourth hydroxyl group [6] and synthesize tetrakis-substituted product with the following structure:

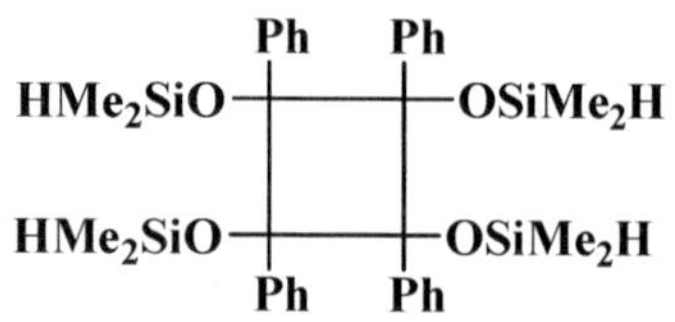

V

Compound (5) of structure V was also synthesized by a single-stage method by HFC reaction of tetrole with dimethylchlorosilane under rigid conditions with substitution of ether by toluene or benzene at the final stage.

To synthesize organocyclosiloxanes with chlorine functional groups at atoms of silicon, HFC of tetrole with dimethylchlorosilane proceeding under soft conditions at 1:4.0 and above 1:5.0 ratios of initial components was studied [7]. The reaction proceeded in the presence of acceptor – pyridine, in anhyd-rous ether in the temperature range of 20 - 30°C. At the final stage, the reaction mixture was boiled. Despite the excess of dimethylchlorosilane, substitution of all four hydroxyl groups by dimethylchlo-rosiloxy-groups under the mentioned conditions has failed. By analogy, the reaction with methyl-phenylchlorosilane proceeds. In the case of diphenyldichlorosilane, the reaction proceeds slowly and under more rigid conditions. Thus, HFC of dimethyldichlorosilane with methylphenyldichlorosilane produces bi- and tetracyclic compounds, proceeding in accordance with the following scheme [7,8]:

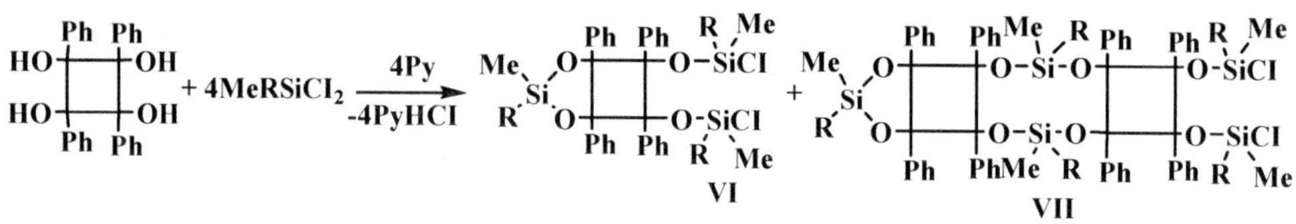

Scheme 4.

To prove the structure of bicyclic compounds, hydrolytic polycondensation of the compound of structure VI [R = Me (6), Ph (8), Table 1] in ether solution was performed [7,8]. It is found that synthesis of organotricyclononasiloxane at the hydrolysis stage in diluted solution in the absence of hydrogen chloride acceptor does not proceed, because cyclization of this type requires more rigid conditions. Tricyclononasiloxanes of structure IX (compounds 10 and 11) were synthesized only after boiling of the products in 5% diluted benzene solution during several hours with the Dyne-Stark trap, at fractiona-tion of the reaction products. The reaction proceeds as follows:

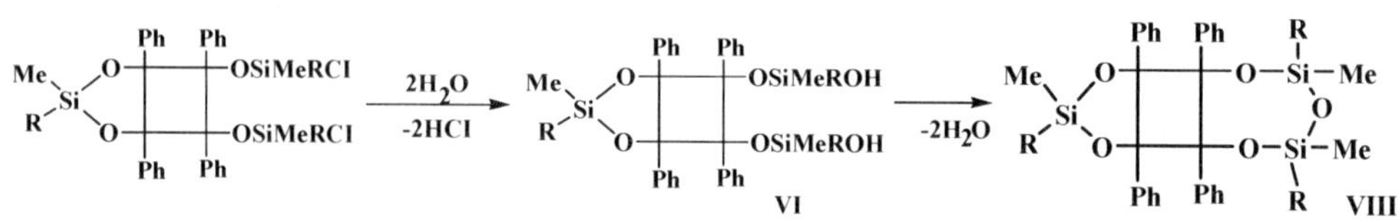

Scheme 5.

where R = Me (10), Ph (11).

To obtain bis(diethylamino)bicyclosiloxanes, aminolytic splitting of *cis*-tetrole reaction products by di-organodichlorosilanes was performed [9-10].

Diethyl amine was used as hydrogen chloride acceptor. The reaction proceeds as follows:

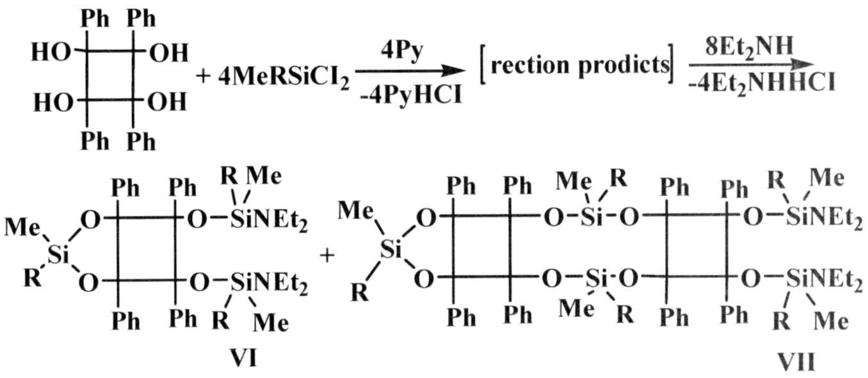

Scheme 6.

where R = Me, Ph.

It should be noted that in the case of phenyl framing, the compound of structure VII was not aminolytically split.

Under soft conditions, similar to scheme (4), HFC of tetrole with silicon tetrachloride at 1:4 ratio of initial components proceeds. However, the product of intermolecular condensation differs from the compound of structure VII. The following reaction scheme was suggested [9]:

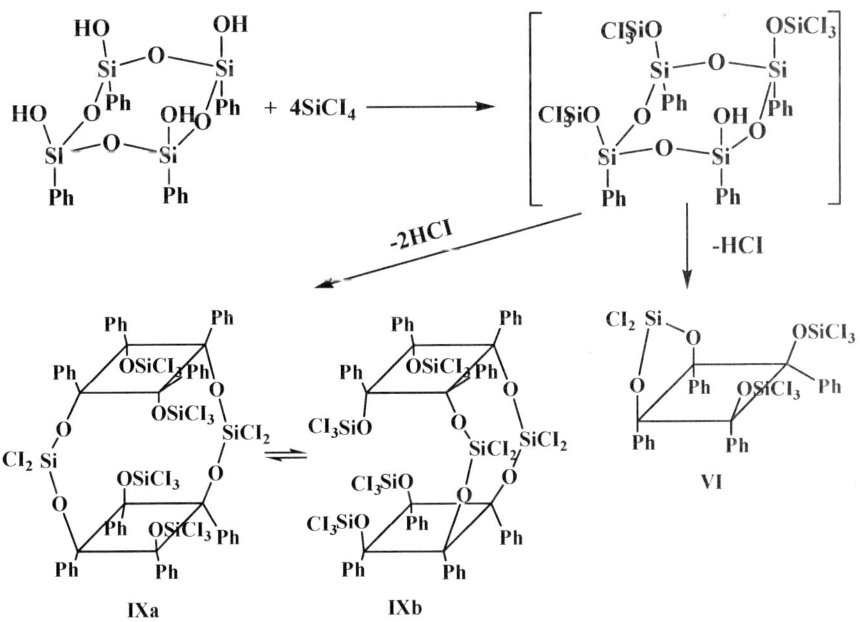

Scheme 7.

The authors suggest the presence of isomeric structures IXa and IXb in the reaction (7). Polycyclic compounds possessing no trichlorosiloxy-group were synthesized in HFC reaction of tetrole with double excess of silicon tetrachloride. The reaction proceeds in accordance with the following scheme:

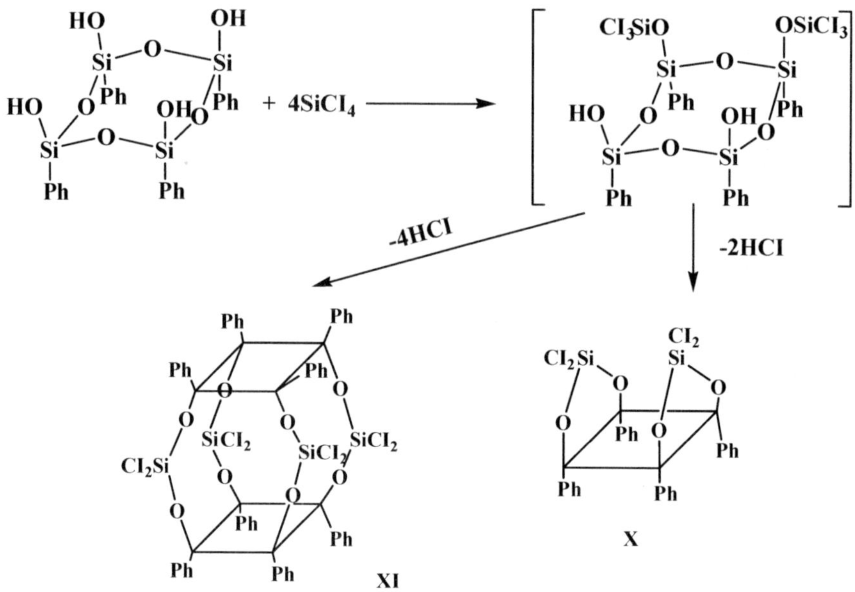

XI

X

Scheme 8.

HFC reaction of tetrole with dimethyldichlorosilane was studied at the ratio of initial components 1:5 in diluted diethyl ether (solution concentration $C = 0.033$ g/ml) at 20 - 25°C, with future diluter substi-tution by benzene and temperature increase to 80°C. Fractionation of the reaction products gave tetrole derivatives, in which all four groups are substituted by chlorodimethylsiloxy-groups, and intermolecu-lar condensation product. The reaction proceeds as follows:

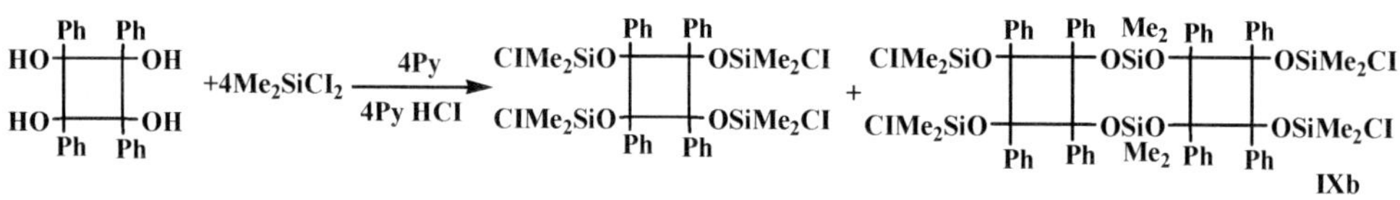

IXb

Scheme 9.

Composition and structure of the compounds, synthesized by the scheme (9), are determined on the basis of ultimate and functional analysis, determination of the molecular mass, and from IR and NMR spectra.

The influence of partial substitution of organochlorosilanes by 1,3-dichlorotetraorgano-disiloxanes(at tetrole interaction with organochlorosilanes) on intra- and intermolecular cyclization, proceeding in the intermediate reaction products, is studied [9, 11].

Benched condensation of tetrole with 1,3-dichlorotetraorganodisiloxane and organochlorosilanes proceeding in diluted solutions in the presence of hydrogen chloride acceptor in the temperature range of 0 - 35°C may form organobicyclononasiloxanes of structure XII, organotricyclodecasiloxanes of structure XIII, organobicyclohexasiloxane of structure VI and products of intermolecular condensation as the final products. The compounds of structures VI and XIII may be synthesized by tetrole inter-action with the single reagent, i.e. by tetrole interaction with organochlorosilane or chlororgano-disilo-xane. General scheme of the reaction is the following:

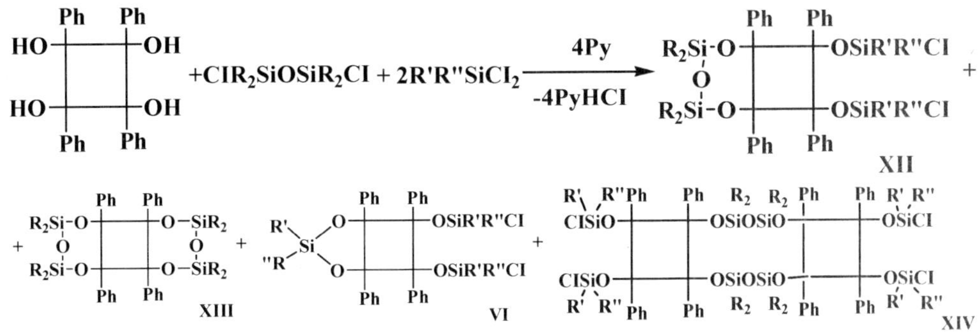

Scheme 10.

where R = Me, Ph; R' = R" = Me, CI; R' = Me, R" = Ph, Vin, CI.

Possible formation of polycyclic compounds of structures VI, VII and XIII is limited by introduction of 1,3-dichlorotetraorganodisiloxanes at the initial stage of the reaction. Negligible amounts of the com-pound of structure XIII can be formed at the interaction of tetrole with 1,3-dichlorotetraorganodisilo-xane, but it is easily fractionated from the reaction products due to its solubility in used solvents at very high dilution degree only.

The reaction scheme (10) indicates that intermolecular condensation of tetrole with 1,3-dichlorotetra-organodisiloxane and organochlorosilanes proceeds through a tetraorganodisiloxane bridge. However, the authors do not exclude formation of isomeric structures XIV.

To prove the structure XIII, i.e. the fact that benched HFC reaction of tetrole with 1,3-dichlorotetra-organodisiloxanes generally proceeds by 1,3-positions of hydroxyl groups, the authors have performed full hydrolysis of dichlorobicycloheptasiloxane. As a result, corresponded dihydroxy-derivatives were synthesized, boiling of which in benzene solution induced further condensation proceeding with water extraction, collected by Dyne-Stark device. Fractionation of thermal condensation products gave com-pounds of structure XIII with both symmetrical and asymmetrical disposition of substituting agents. The reaction proceeds as follows:

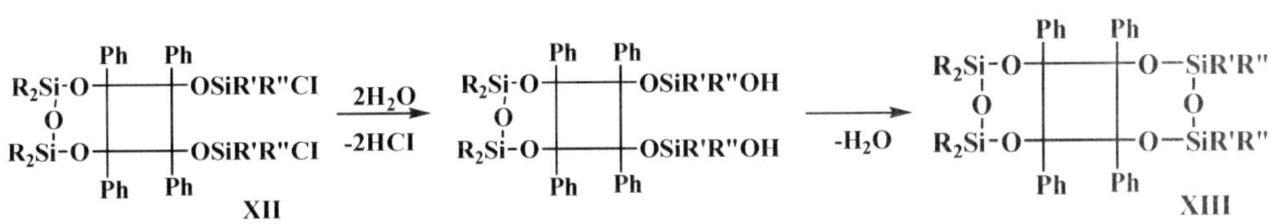

Scheme 11.

where R = R' = R" = Me; R = R' = Me, R" = Vin; R = Ph, R' = R" = Me.

By analogy, full hydrolysis of tetrachlorobicyclononasiloxane under soft conditions was performed, and intramolecular cyclization type of its proceeding with formation of dihydroxytricyclodecasiloxane by the following scheme was proved:

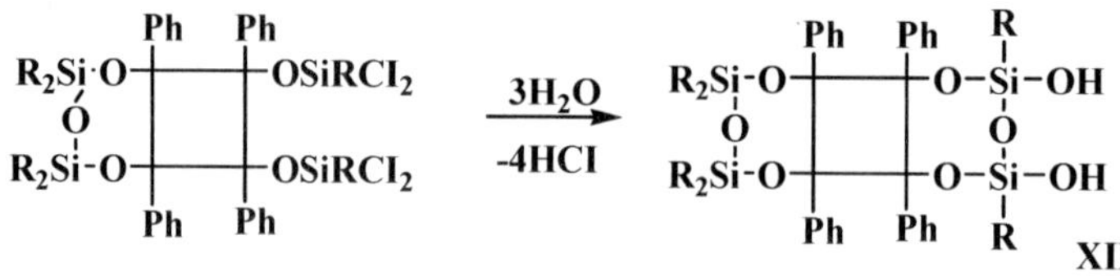

XIII

Scheme 12.

where R = Me, Ph.

By partial hydrolysis of tetrachlorobicyclononasiloxane of structure XII under soft conditions dichloro-bicyclodecasiloxanes were synthesized by the scheme as follows:

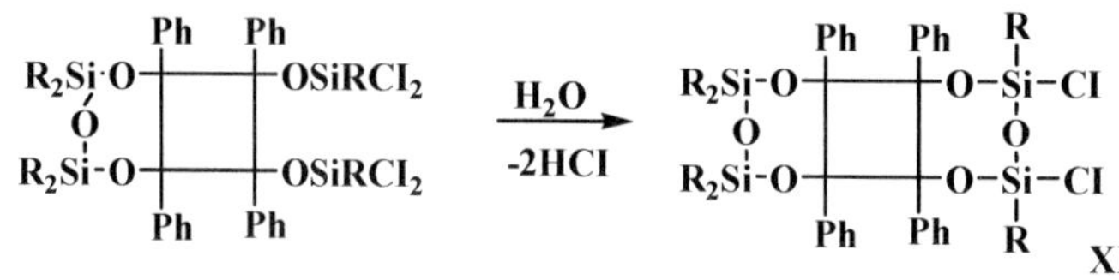

XIII

Scheme 13.

where R = Me, Ph.

Composition and structure of the products, synthesized in the benched heterofunctional polycondensation, and the ones of their transformation were determined by functional and ultimate analysis, molecular mass measurement, and from IR and ^{1}H NMR spectral data.

As far as the above-mentioned results indicate predominant proceeding of tetrole condensation with 1,3-dichlorotetraorganodiciloxane by 1,3-position of hydroxyl groups, further on, the benched HFC re-action of tetrole with 1,3-dichlorotetraphenyldisiloxane and 1,1,3,3-tetrachloro-1,3-diphenyldisiloxane was studied [9,10]. The sequence of introduction of haloid-derivatives of phenylsiloxanes plays the im-portant role, because fractionated condensation products form three-dimensional systems in the excess of 1,1,3,3-tetrachloro-1,3-diphenyldisiloxane. That is why, implementation of the reaction by suggested scheme (13) requires introduction of 0.05 – 0.10 mole excessive 1,3-dichlorotetraphenyldisiloxane, recalculated for tetrole, and a half of pyridine into the tetrole solution, first, and then 1,1,3,3-tetra-chloro-1,3-diphenyldisiloxane and the rest of pyridine. The reaction proceeds by the scheme as follows:

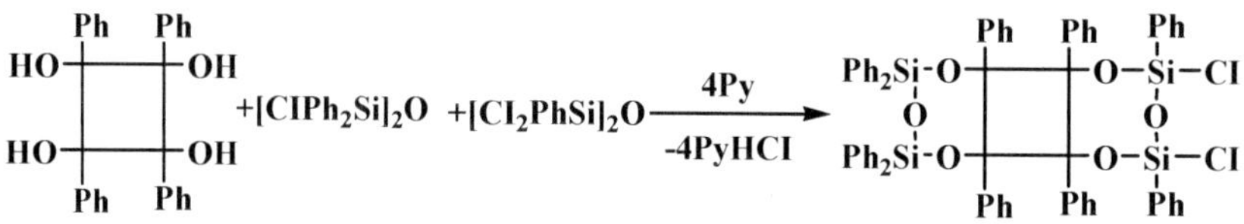

XIII

Scheme 14.

Synthesized compound (34) (structure XIII, Table 1) was identified on the basis of ultimate and functional analysis, molecular mass measurement and IR spectral data. Structure of the compound (34) was determined by hydrolytic condensation. Future processing of hydroxyl derivative by trimethyl-chlorosilane gave an opportunity to separate a compound with trimethylsiloxy-groups in accordance with the scheme as follows:

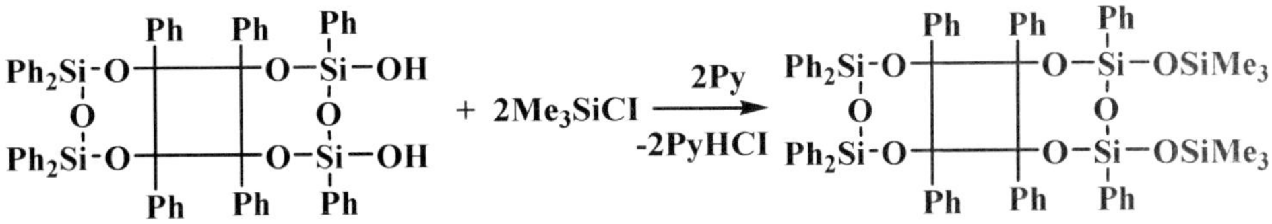

XIII

Scheme 15.

The final conclusion about the structure of decaphenyltricyclosiloxane was made on the basis of ^{1}H NMR spectrum [8, 10]. The influence of temperature of tetrole condensation with organochlorosilanes on the structure and yield of compounds formed were studied in detail in these works.

1,3-Divinylorganotricyclodecasiloxane was synthesized by benched HFC reaction of tetrole with 1,3-dichlorotetramethyldisiloxane and 1,3-dichloro-1,3-divinyl-1,3-dimethyldisiloxane at 1:1:1 ratio of initial components in the presence of pyridine by the scheme as follows [12]:

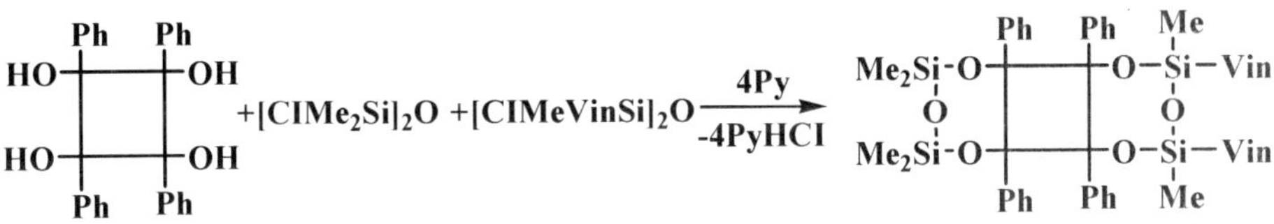

XIII

Scheme 16.

Physical and chemical parameters of the compound, synthesized by the reaction (16), coincide with these of the compound, synthesized by the reaction (11).

To obtain divinyltricyclohexasiloxane, the authors have studied HFC reaction of *cis*-tetrole with me-thylvinyldichlorosilane at 1:2 ratio of initial components in the presence of acceptor, triethylamine. Op-timal conditions of obtaining 1,7-divinyl-1,7-dimethyl-3,5,9,11-tetraphenyltricyclohexasiloxane were selected, which proceeds in accordance with the following scheme [13]:

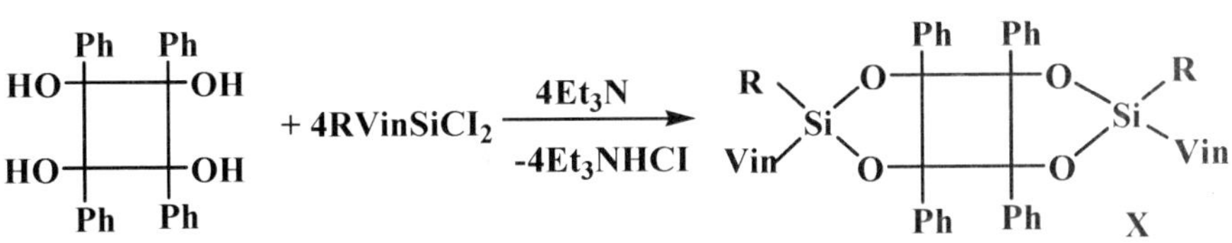

X

Scheme 17.

Where R = Me, Ph.

^{1}H NMR spectroscopy method has indicated that the compound, synthesized by the reaction (17) (for R = Vin), displays three isomeric shapes: A, B and C, occurrence of which is associated with different dispositions of methyl and phenyl groups:

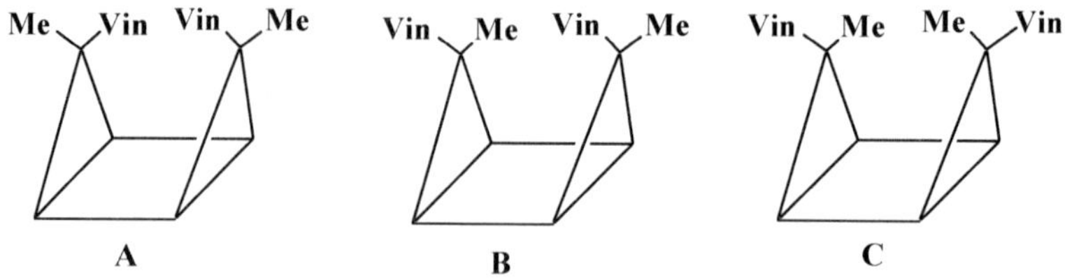

By positions in stereo-isomers, methyl groups are related to three nonequivalent groups that explains the presence of three reflexes from methyl protons in NMR spectra of divinyltricyclohexasiloxane. Analysis of reflex intensities in NMR spectra of separate fractions, obtained by fraction recrystallization of the reaction (17) product, by the method of selecting all possible alternatives allowed relation of each of three reflexes to a definite group of methyl protons. The only alternative suits experimental spectra: singlet with the shift at δ=0.38 ppm (A), δ=0.41 ppm (B) and δ =0.51 ppm (C). Based on this alternative, concentration of each of three stereo-isomers in the mixture is calculated from the reflex intensities: A–18%, B–57% and C–25%.

Recently, it was shown that hydrolysis of 1,1,5,5-tetrachlorotetraphenylcyclotetrasiloxane produces 1,1,5,5-tetrahydroxytetraphenylcyclotetrasiloxane (T-2) [14], which is *cis*-tetrole (T-1) isomer [15]. It was found that contrary to T-1, T-2 forms an aggregate with pyridine in accordance with the following scheme:

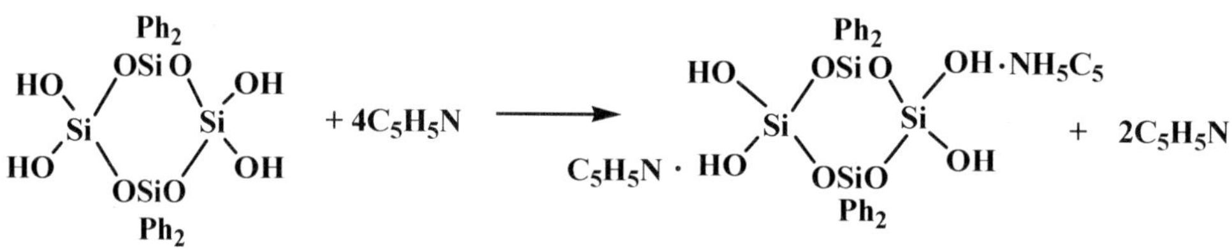

Scheme 18.

White crystallites of T-2 complex with pyridine display low solubility in low-polar organic solvents and good solubility in polar ones. It is shown in the literature that tetraphenyldisiloxane diol and hexaphenyltrisiloxane diols form complexes with organic bases [16,17]. Interaction of dipyridine-tetrahydroxytetraphenylcyclosiloxane [18] with excessive dimethyldichlorosilane in the presence of py-ridine forms a difunctional bicyclic compound in accordance with the following scheme:

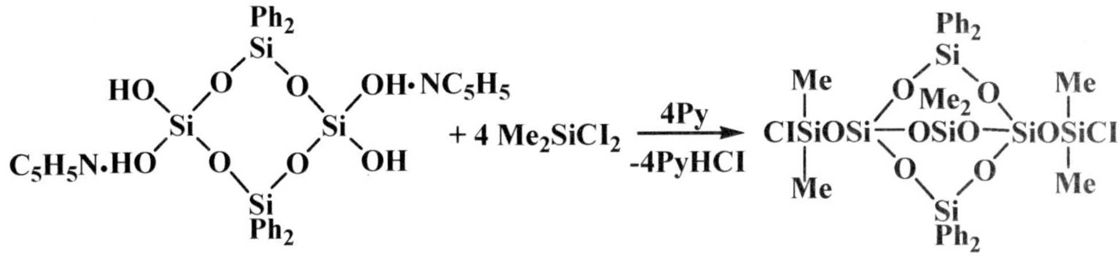

XVI

Scheme 19.

By analogy, condensation of the complex with excessive 1,3-dichlorotetramethyldisiloxane in the presence of pyridine proceeds:

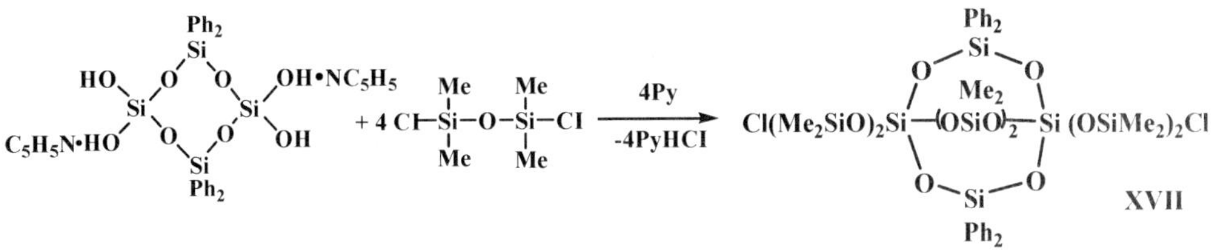

Scheme 20.

The interaction of T-2 with diorganodichlorosilanes in the presence of acceptor (pyridine) proceeds differently, giving tetra functional branched organocyclosiloxanes in accordance with the following scheme [19]:

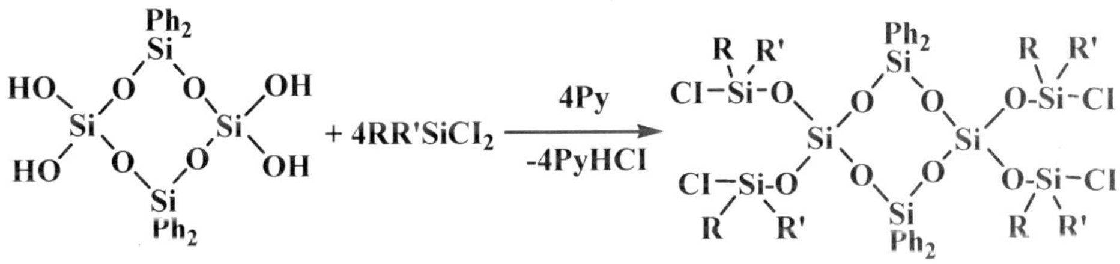

XVIII

Scheme 21.

where R=R'=Me, Ph, Et. R≠R'.

This gives grounds to assume that dipyridinetetrahydroxytetraphenylcyclotetrasiloxane possesses the *cis*-structure. As suggested, formation of difunctional compounds proves [18] that free and linked hydroxyl groups in T-2 complex with pyridine differ by reactivity in relation to organochlorosilanes. Due to high basicity of oxygen in hydroxyl groups, HFC reaction rate of ≡Si-OH·NH_5C_5 group with organochlorosilanes is higher than that of ≡Si-OH group.

Hydrolysis of tetrakis(chlorodiorganosiloxy)tetraphenylcyclotetrasiloxane (structure XVIII) with water excess at 5-10°C has produced appropriate tetrahydroxy-derivatives of the following structure [9,10]:

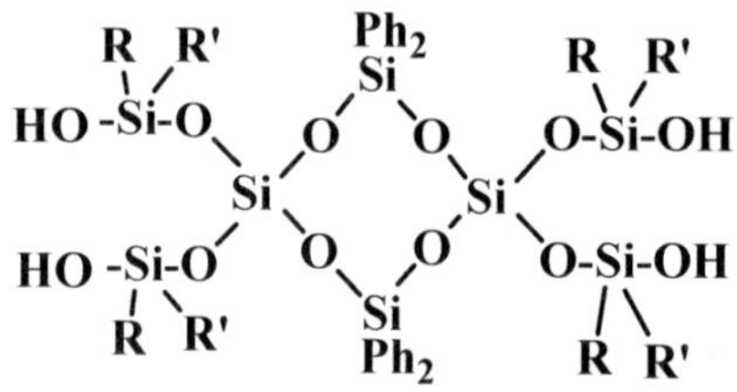

XVIII

where R=R'=Me, Ph; R≠R'.

Due to comparatively high reactivity of hydroxyl groups, which are easily condensed, the yield of hydroxyl derivative with methyl radicals (R=R'=Me) at the atom of silicon from hydrolysis proceeding in the presence of acceptor is negligible, and without acceptor was not obtained at all. On the contrary, stability of hydroxyl groups of the compound (R=R'=Ph) promoted synthesis of this product with high yield in hydrolysis of the compound both in the presence and in the absence of HCI acceptor.

HFC reaction of tetrakis(hydroxydiorganosiloxy)tetraphenylcyclotetrasiloxane (structure XVIII) with organotrichlorosilanes at 1:2 ratio of initial components was performed at 15-20°C in organic solvent, in the presence of HCI acceptor. Benzene, toluene, diethyl ether, tetrahydrofuran, methyl ethyl ketone or mixture of the solvents was used. Dichlorospirotricyclosiloxanes were synthesized in the presence of pyridine in accordance with the scheme as follows [18, 19]:

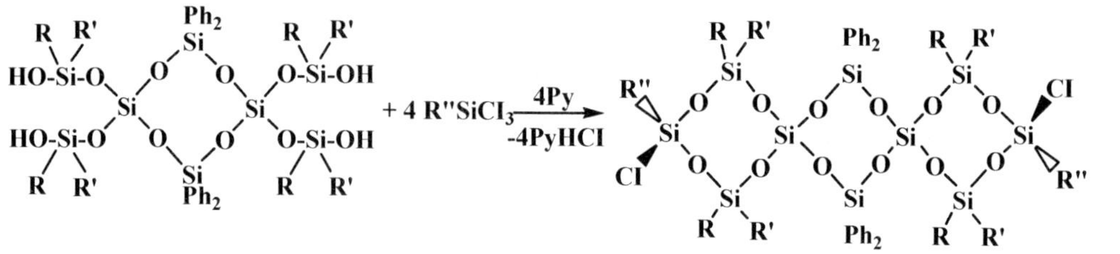

XIX

Scheme 22.

where R=R'=Ph; R"=Me, Ph.

By hydrolysis of dichlororganospirotricyclosiloxane, appropriate dihydroxy-derivatives were synthe-sized in accordance with the following scheme:

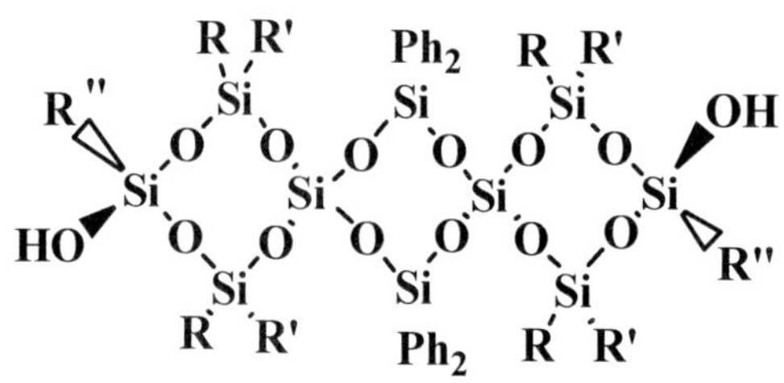

XIX

where R=R'=Me,Ph; R≠R'.

Composition and structure of synthesized difunctional spirotricyclosiloxanes were determined by func-tional and ultimate analysis, molecular mass measurement, and from IR and ^{1}H NMR spectral data. As indicated, found ratios of protons of methyl and phenyl groups correlate well with calculated values.

For the purpose of synthesizing arylenecyclosiloxane blocks with functional groups at silicon atom, the reaction was performed in two stages. At the first stage, HFC of 1,4-bis(dichlorophenylsilyl)benzene with dihydroxydiphenylsilane at 1:4 ratio of initial components in the presence of pyridine in the tem-perature range of -5÷-10°C was performed. After recrystallization of the reaction product from tolu-ene–heptane solution 52% yield of tetrakis(hydroxydiphenylsiloxy)-1,4-bis(phenylsilyl)benzene was obtained. The reaction proceeds in accordance with the following scheme [20]:

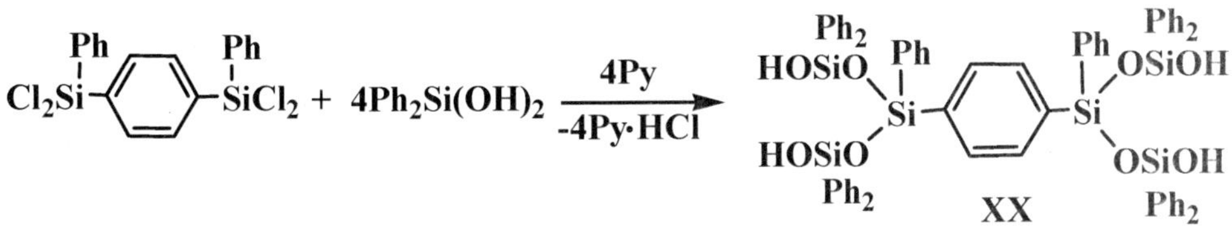

Scheme 23.

At the second stage, HFC of the product of structure XX (compound 46, Table 1) with organo-trichlorosilane at 1:2 ratio of initial components, in the presence of pyridine and temperature range of -5÷-100C was carried out and dichloroarylenecyclosiloxanes were obtained. In this case, the reaction generally proceeds by intramolecular cyclization mechanism in accordance with the following scheme:

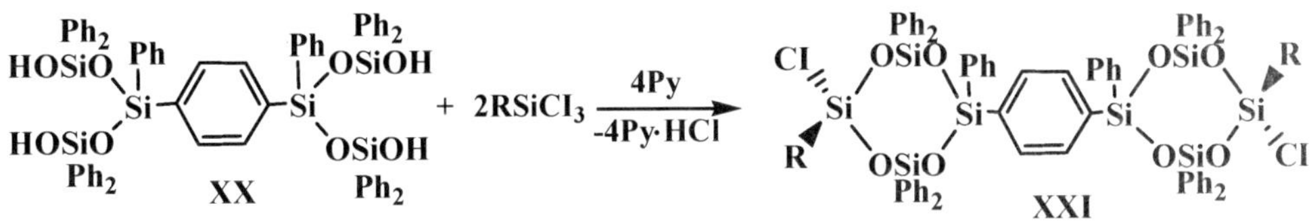

Scheme 24.

where R = Me, Ph.

Compounds of structure XXI (R = Me (47), Ph (48), Table 1) were also synthesized by a single stage HFC of 1,4-bis(dichlorophenylsilyl)benzene with dihydroxydiphenylsilane and organotrichlorosilanes at 1:4:2 ratio of initial components in the presence of pyridine and temperature range of –5 ÷10°C, and the yield of compounds was below ~32 %.

Hydrolysis of dichloroarylenecyclosiloxanes of structure XXI, proceeding under soft conditions in the presence of sodium bicarbonate and further recrystallization from toluene-heptane solution produced appropriate dihydroxy-compounds as follows:

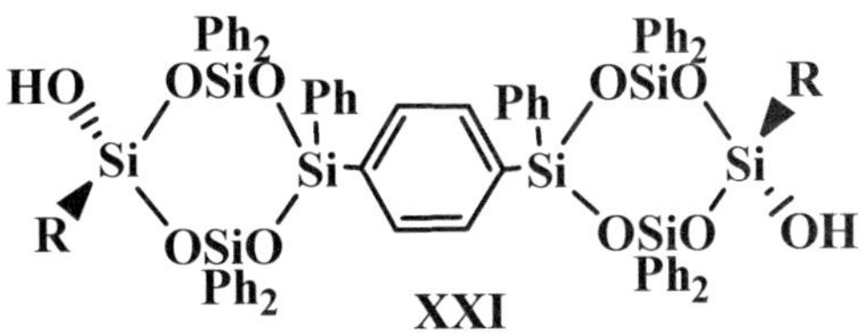

where R = Me (49), Ph (50), Table 1.

Composition and structure of synthesized dichloro- and dihydroxyarylenecyclosiloxanes (46 - 50) of the structure XXI were determined by ultimate analysis, molecular mass measurements, and from IR and ^{1}H NMR spectral data.

^{1}H NMR spectrum of compound 47 (Table 1) displays two singlet reflexes with chemical shifts δ=0.56 and δ=0.52 ppm typical of protons of methyl groups of *cis*- and *trans*-configuration. The ratio of *cis*- and *trans*-configurations equals 60 and 40%, respectively.

Some physical and chemical parameters of all synthesized organobi-, organotri- and organotetra-cyclosiloxanes are shown in Table 1.

Table 1. Some physical and chemical parameters of organobi-, organotri- and organotetra-cyclosiloxanes

№	Compound structure	Yield, %	T_{boil}, ^{0}C T_{melt}, ^{0}C	P, mm.Hg	Refe-rences
1	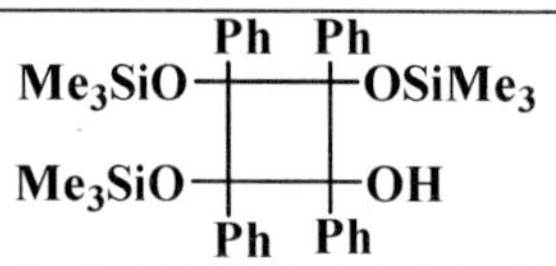	47	225-230	1-2x10^{-3}	[5,8]
2	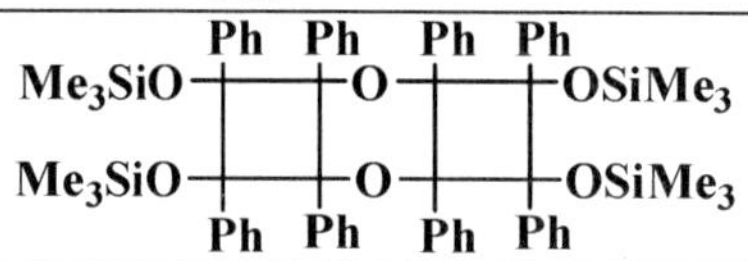	9	318-320	1-2x10^{-3}	[5,8]
3	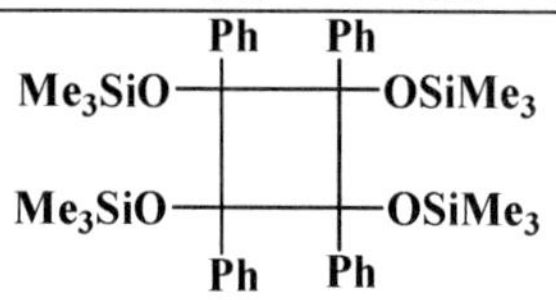	78	241-242	-	[5,8]
4	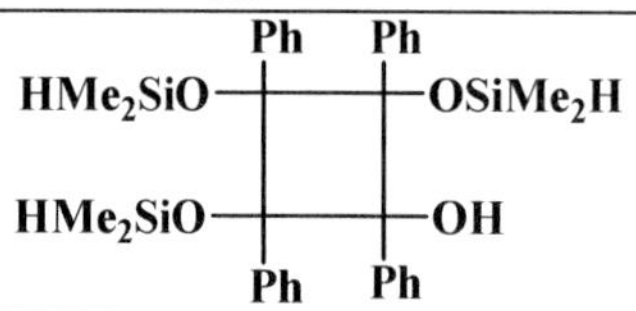	93	190-192	1-2x10^{-2}	[6,10]
5	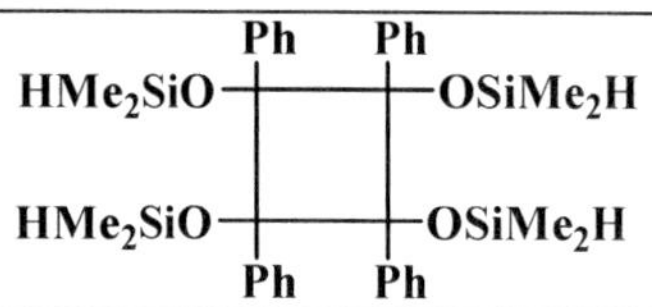	93	215-220	1-2x10^{-2}	[6,10]
6	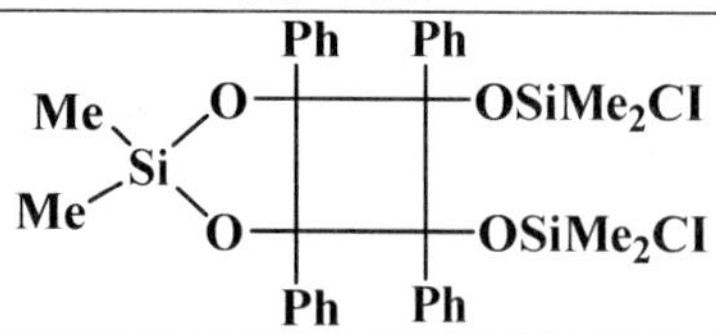	39	197-205	1-2x10^{-2}	[7,8]
7	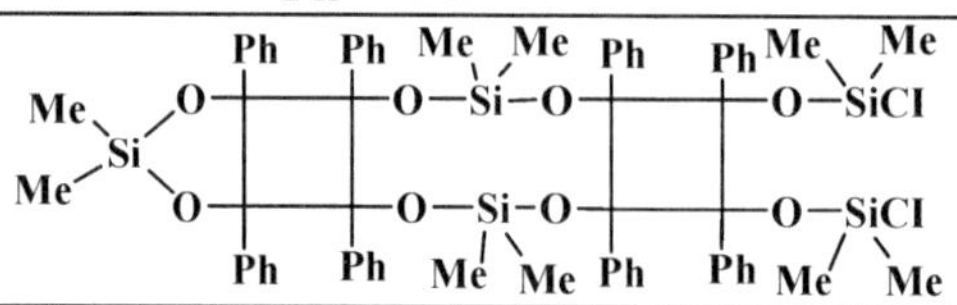	33	290-295	1-2x10^{-3}	[7,8]

Table 1. (Continued)

№	Compound structure	Yield, %	T_{boil}, ^0C T_{melt}, ^0C	P, mm.Hg	References
8	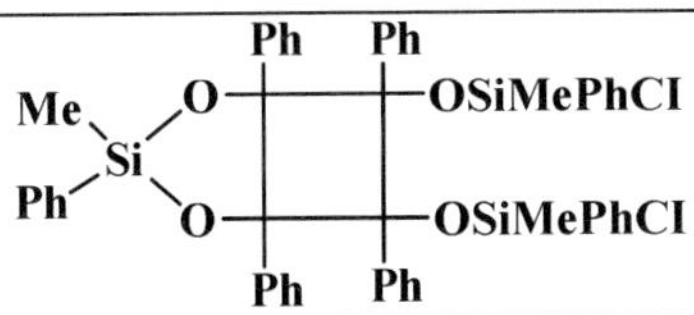	35	322-325	$1\text{-}2x10^{-3}$	[7,8]
9	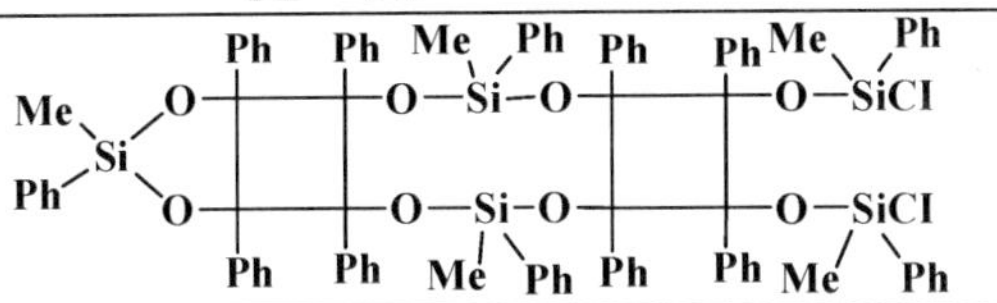	9	405-410	$1\text{-}2x10^{-3}$	[7,8]
10	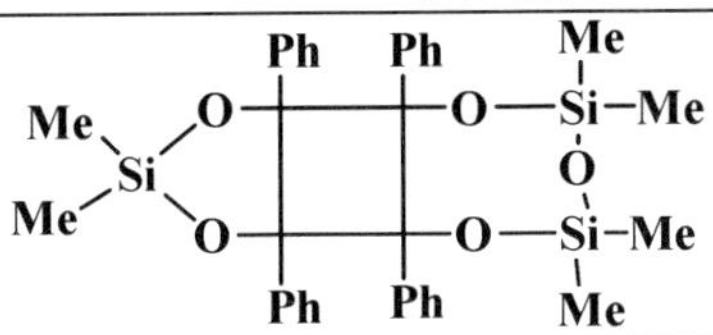	65	232-235	$6\text{-}8x10^{-4}$	[7,8]
11	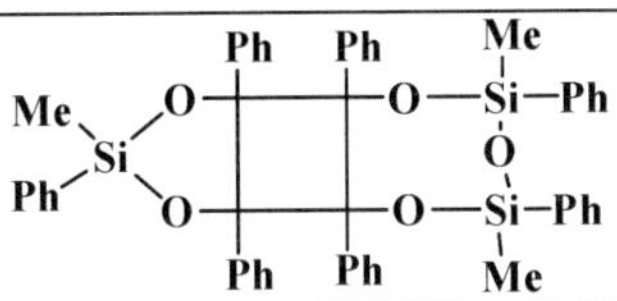	30	310-320	$1\text{-}2x10^{-3}$	[7,8]
12	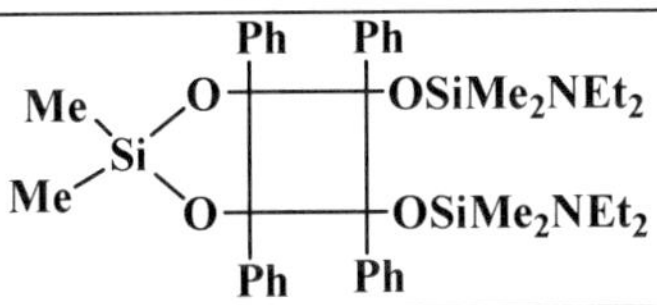	30	222-225	$1\text{-}2x10^{-3}$	[8]
13	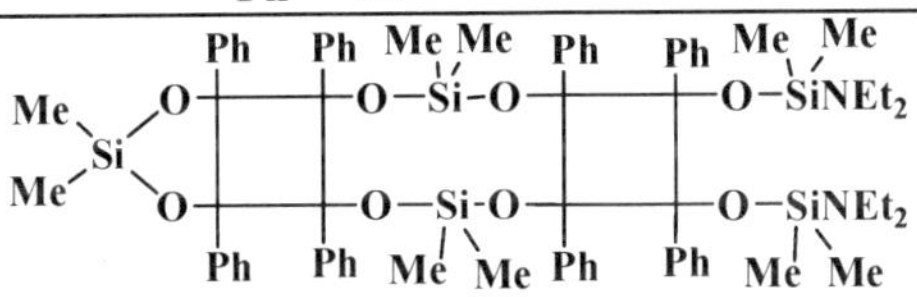	22	320-322	$1\text{-}2x10^{-3}$	[8]
14	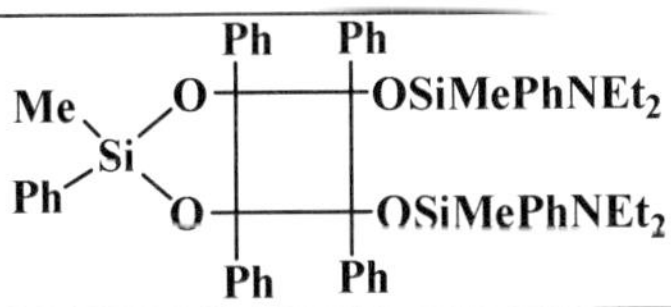	41.5	290-295	$1\text{-}2x10^{-3}$	[8]
15	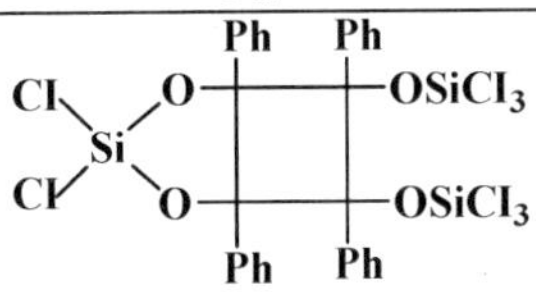				[9]
16	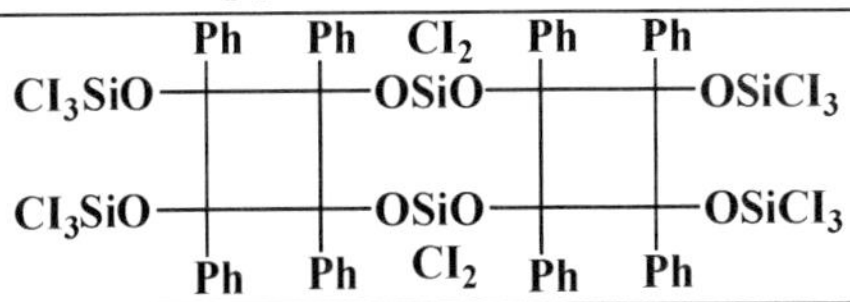				[9]

Table 1. (Continued)

№	Compound structure	Yield, %	T_{boil}, ^0C T_{melt}, ^0C	P, mm.Hg	References
17	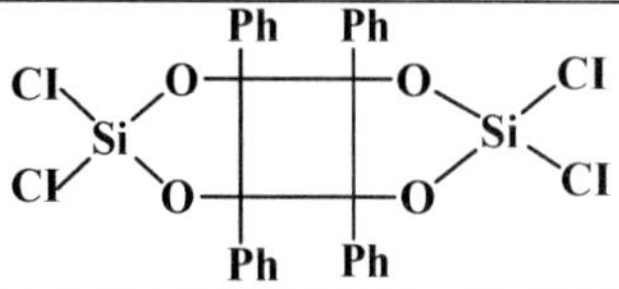				[9]
18	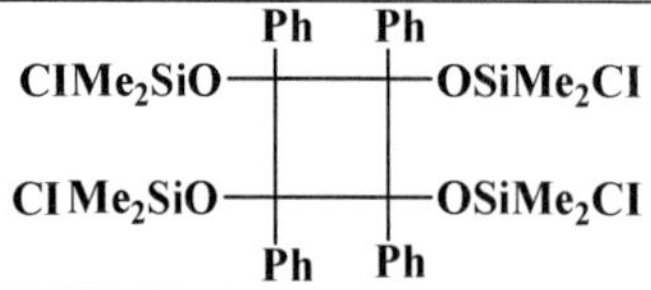	76	232-235	$1\text{-}2x10^{-3}$	[5,8]
19	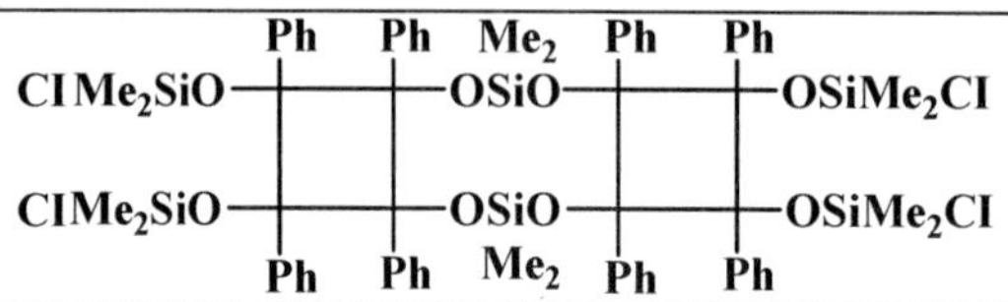	11	320-325	$1\text{-}2x10^{-3}$	[5,8]
20	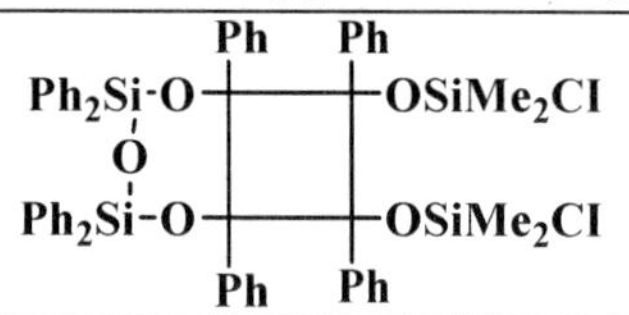	26	310-320	$1\text{-}2x10^{-3}$	[8,11]
21	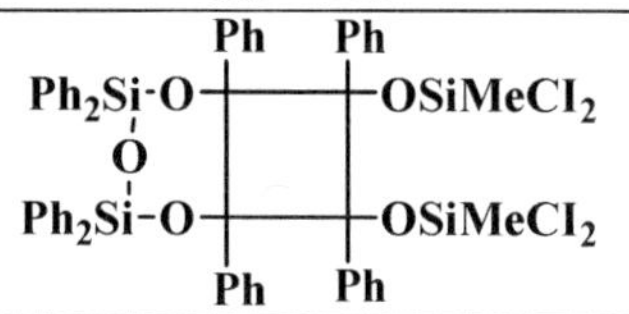	21	315-320	$1\text{-}2x10^{-3}$	[8,11]
22	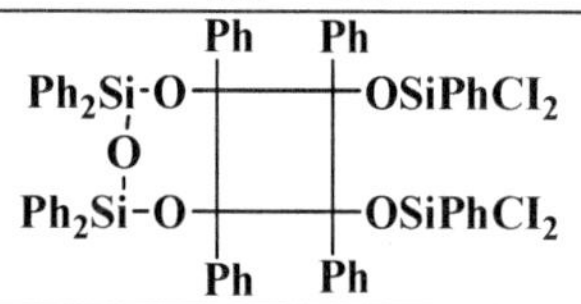	46	345-355	$1\text{-}2x10^{-3}$	[8,11]
23	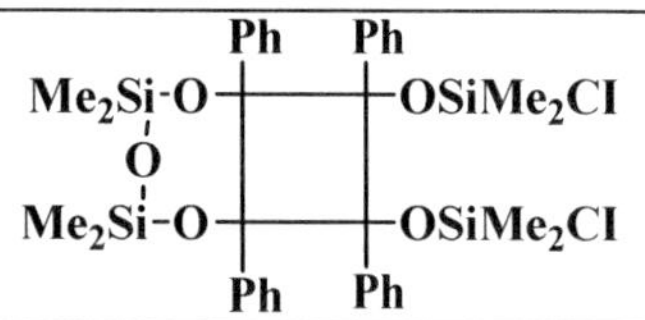	58	220-225	$1\text{-}2x10^{-3}$	[8,11]
24	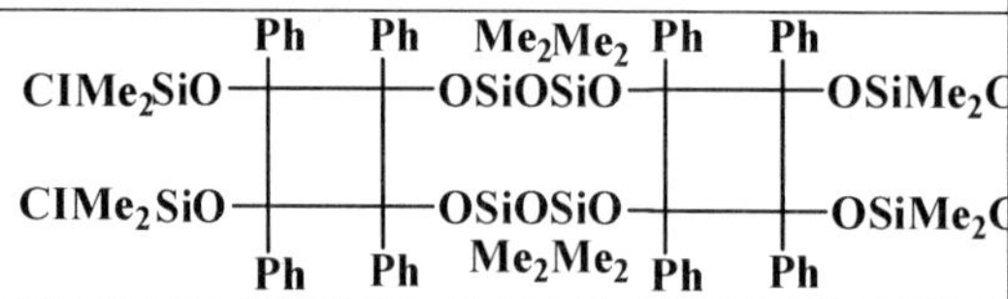	20	348-352	$1\text{-}2x10^{-3}$	[8,11]
25	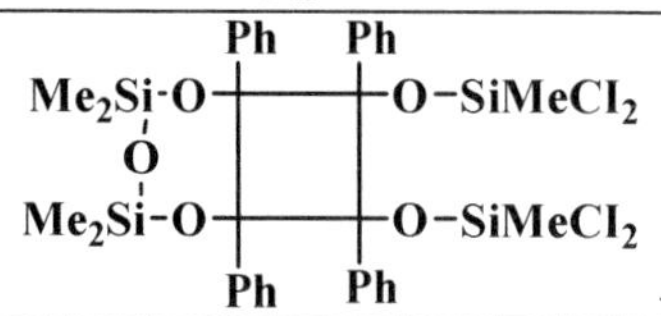	63	226-230	$1\text{-}2x10^{-3}$	[8,11]

Table 1. (Continued)

№	Compound structure	Yield, %	T_{boil}, ^0C T_{melt}, ^0C	P, mm.Hg	Refe-rences
26	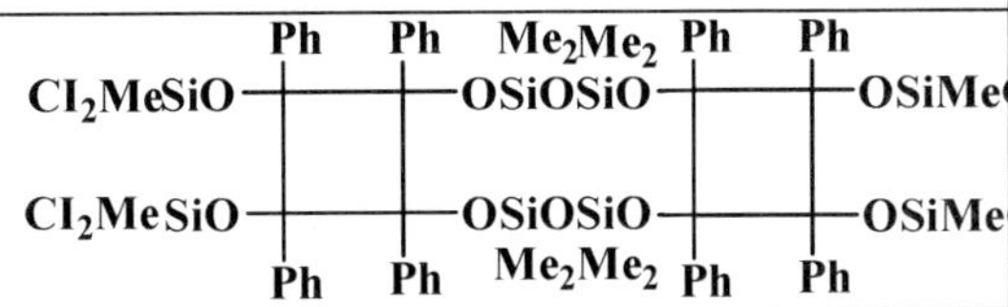	12	310-325	$1\text{-}2x10^{-3}$	[8,11]
27	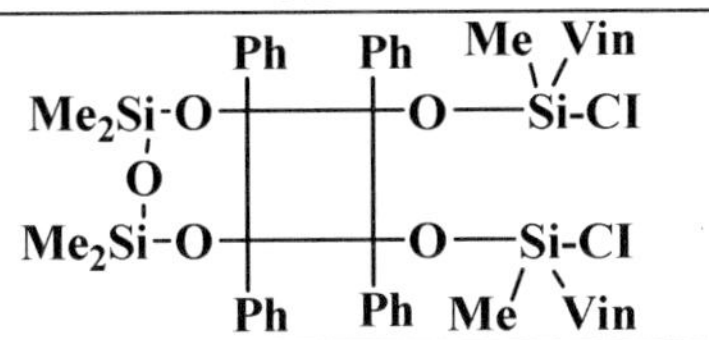	50	235-240	$1\text{-}2x10^{-3}$	[8,11]
28	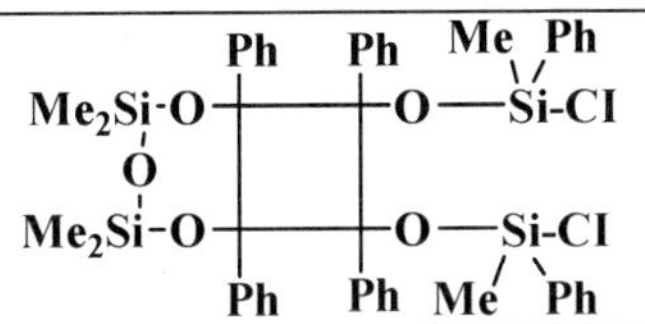	36	295-305	$1\text{-}2x10^{-3}$	[8,11]
29	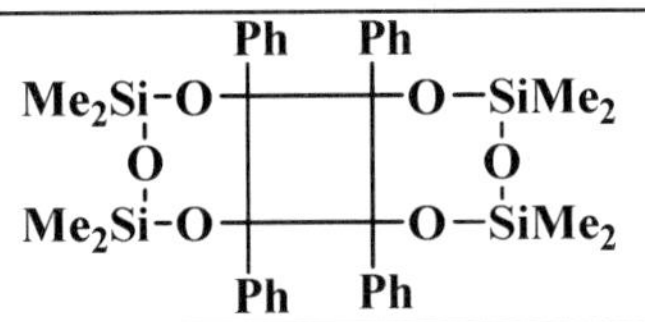	56	235-240	$1\text{-}2x10^{-3}$	[8,11]
30	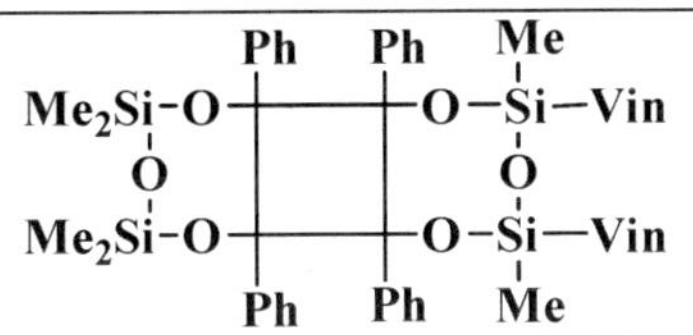	62	245-250	$1\text{-}2x10^{-3}$	[8,11]
31	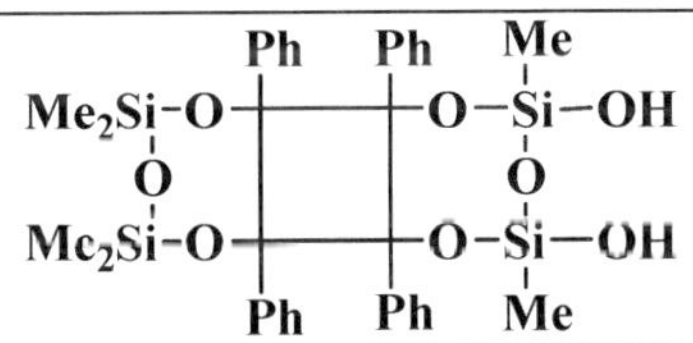	90	75-78	-	[8]
32	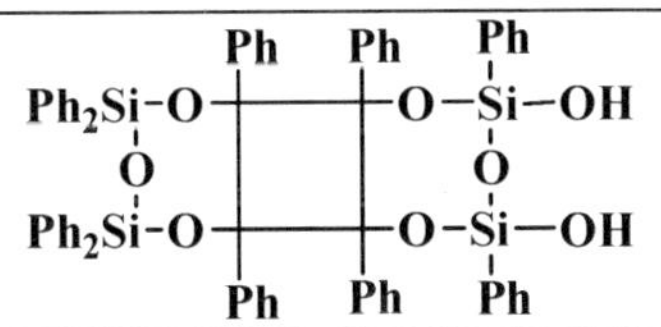	94	110-112	-	[8]
33	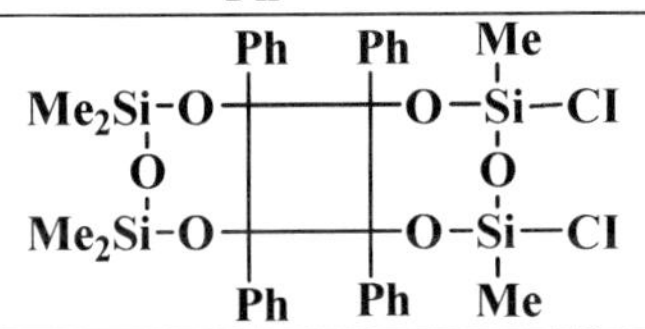	47	240-245	$1\text{-}2x10^{-3}$	[8,11]
34	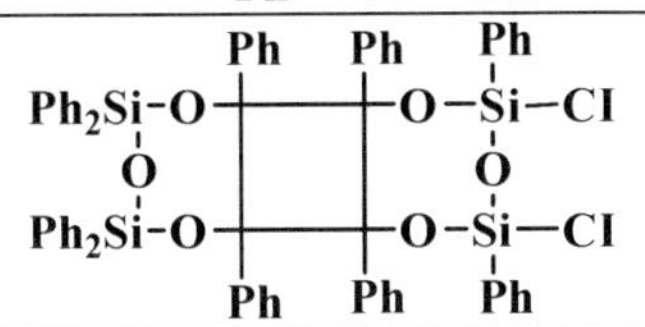	27	350-360	$1\text{-}2x10^{-3}$	[8]

Table 1. (Continued)

№	Compound structure	Yield, %	T_{boil}, ^0C T_{melt}, ^0C	P, mm.Hg	References
35	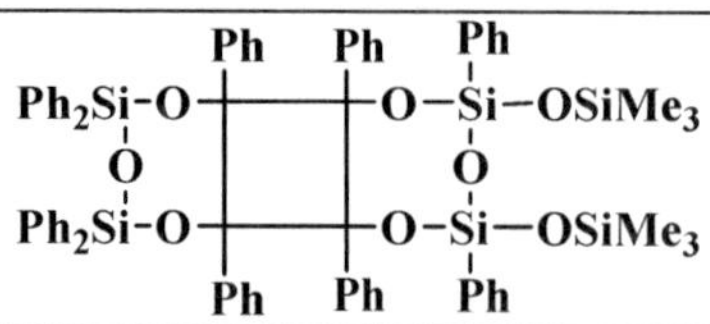	18	320-325	1-2x10^{-3}	[17]
36	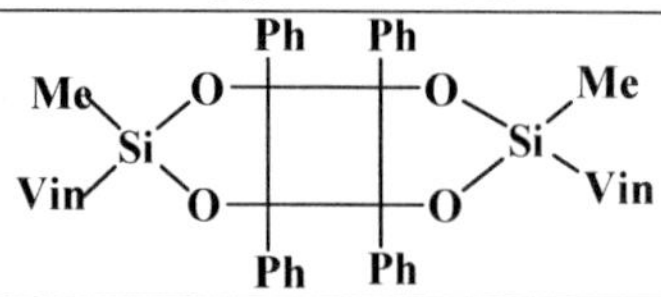	32	- 150-153	-	[13]
37	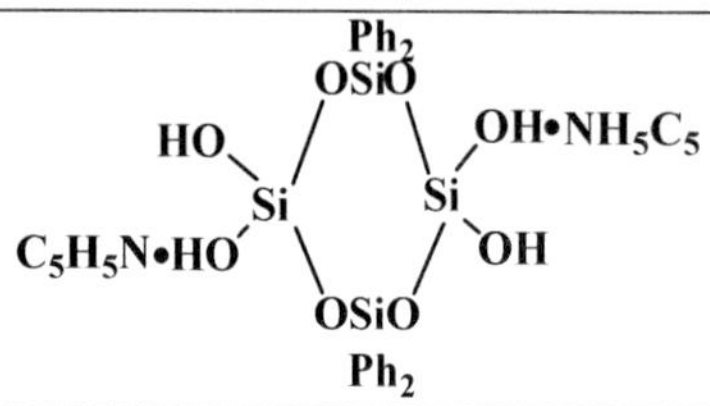	88	118-119	-	[14,17]
38	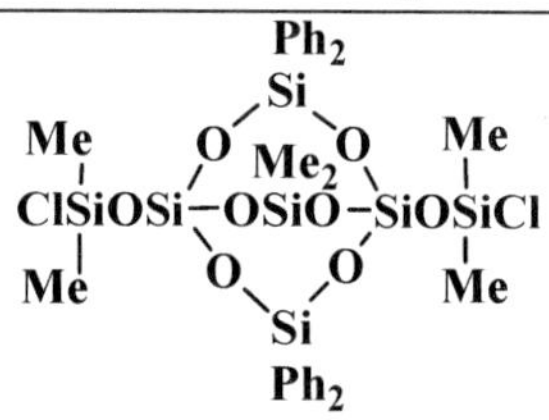	54	212-214	2x10^{-3}	[17]
39	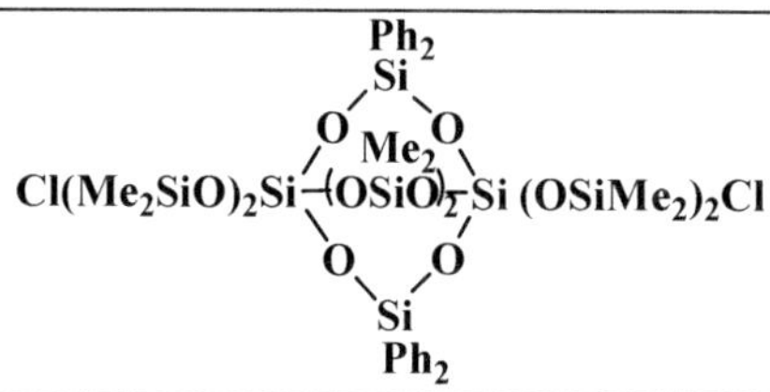	63	229-231	2x10^{-3}	[17]
40	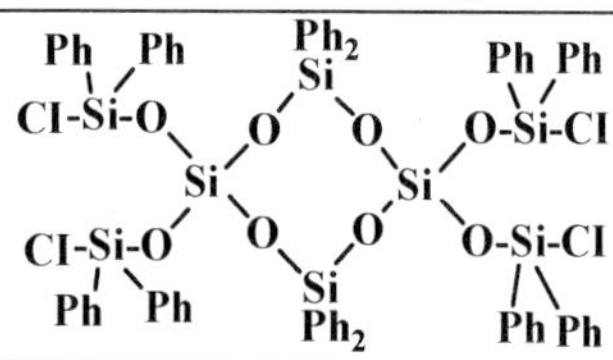	66	150-151	-	[19]
41	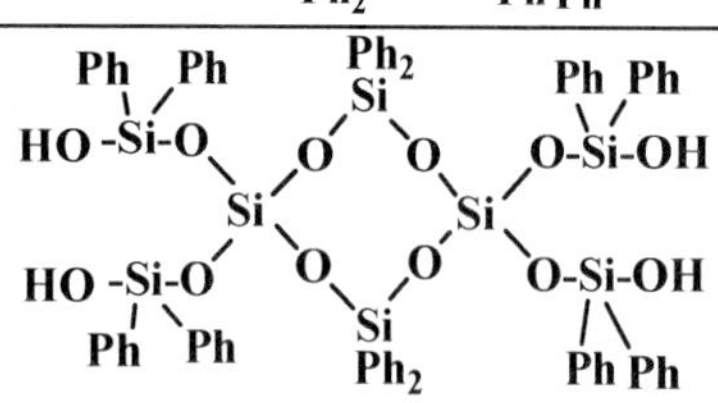	89	179-180	-	[19]

Table 1. (Continued)

№	Compound structure	Yield, %	T_{boil}, ^0C T_{melt}, ^0C	P, mm.Hg	Refe-rences
42	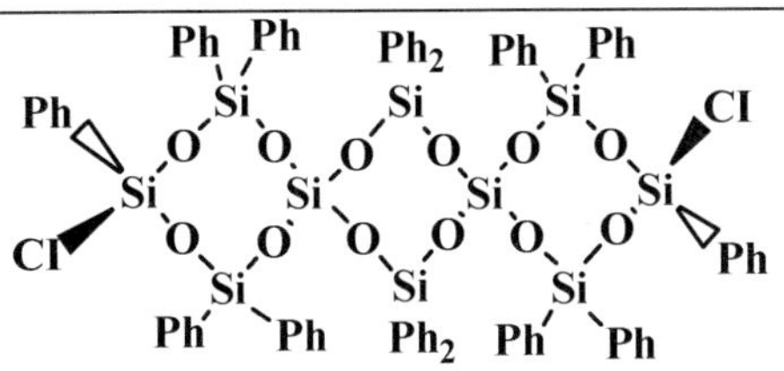	43	198-200	-	[19]
43	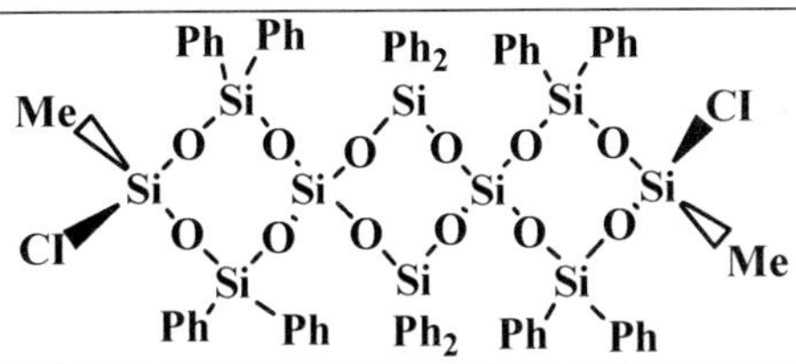	62	139-141	-	[19]
44	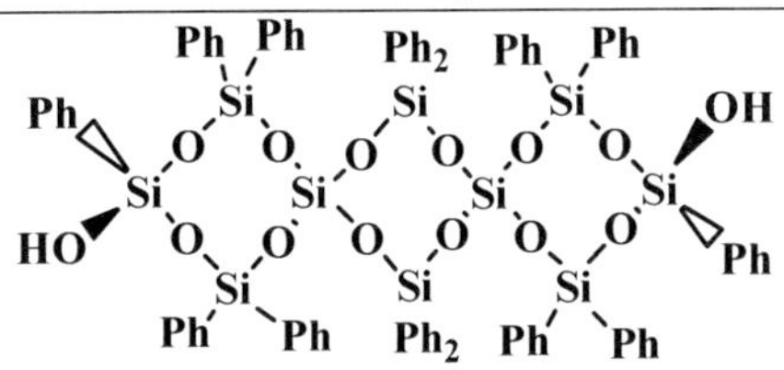	83	216-217	-	[19]
45	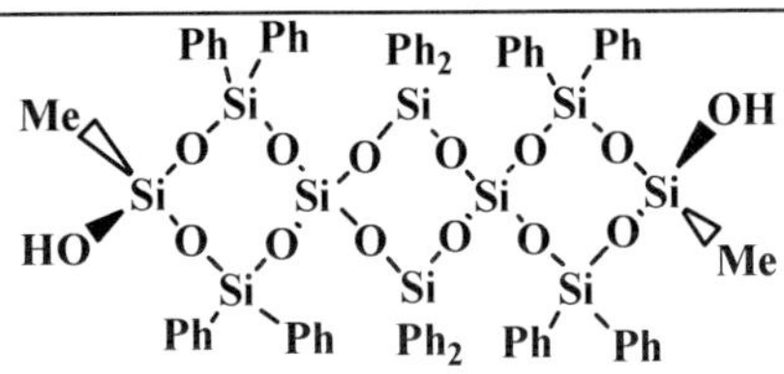	81	153-155	-	[19]
46	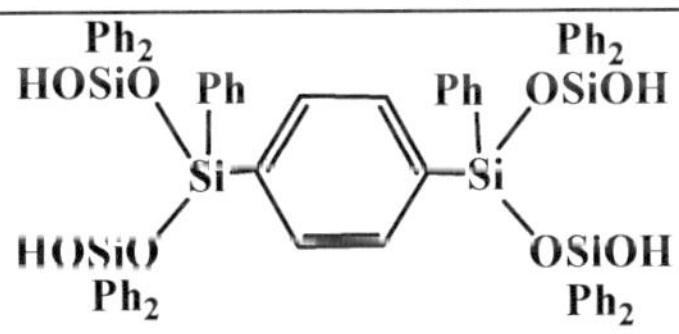	52	- 120-121	-	[20,21]
47	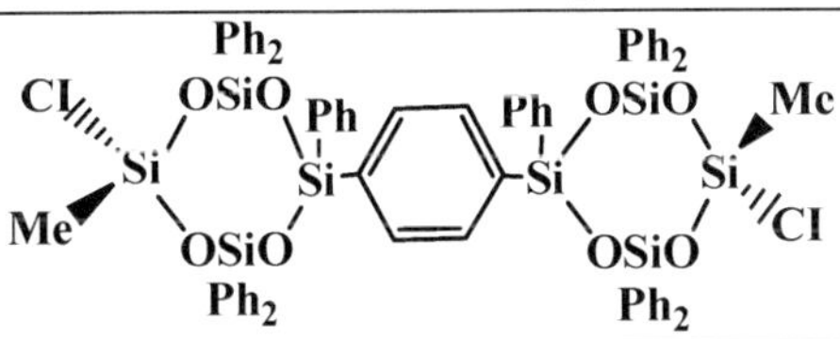	51 (31)	340-345 58-60	1-2x10^{-3}	[20,21]
48	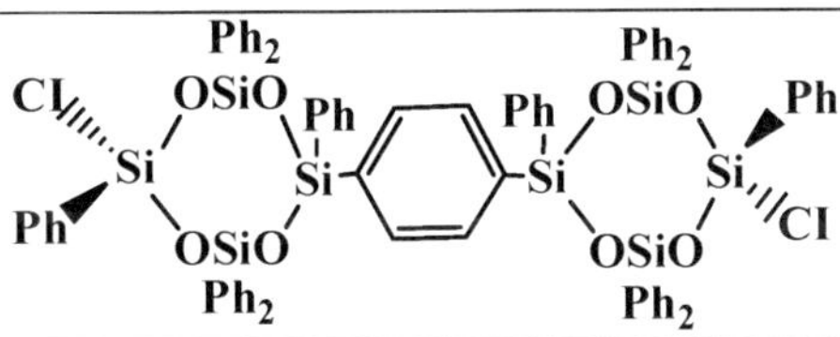	54 (33)	370-380 85-88	1-2x10^{-3}	[25,26]

Table 1. (Continued)

№	Compound structure	Yield, %	T_{boil}, ^0C T_{melt}, ^0C	P, mm.Hg	Refe-rences
49	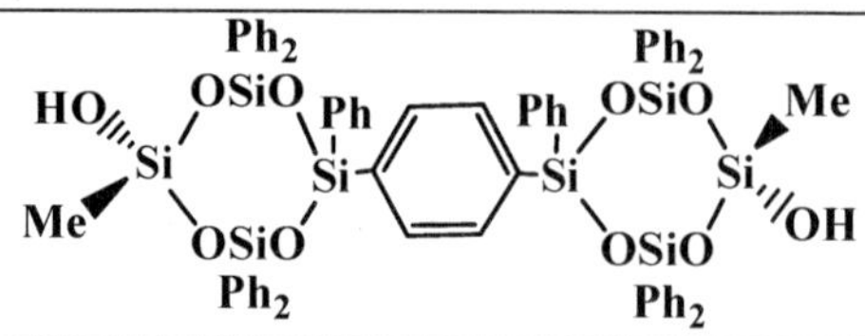	53	- 82-85	-	[20,21]
50	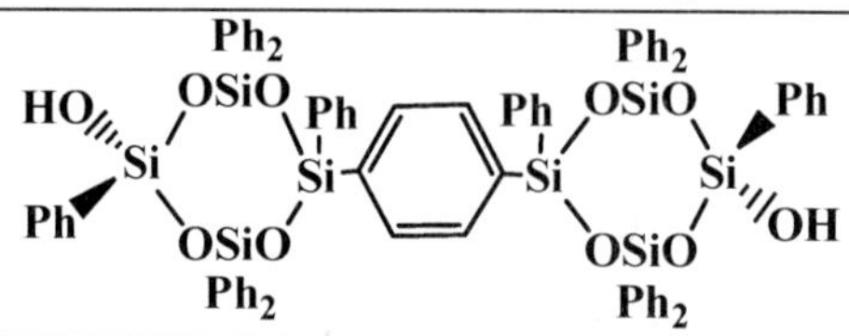	67	- 95-97	-	[20,21]

2. Synthesis of Difunctional Bicyclocarbosiloxanes

For the purpose of synthesizing difunctional bicyclocarbosiloxanes, HFC reaction of 1,2-bis(trich-lorosilyl)ethane with dihydroxydiphenylsilane was studied [21]. The reaction proceeded at 1:2 ratio of initial components in diluted solution of anhydrous toluene and in the presence of pyridine. In the temperature range of -5 ÷ -10°C dichlorobicyclocarbosiloxane with ~44% yield was synthesized. High-er yield of this compound (~69%) was obtained by the two-stage method: at the first stage, tetrachloro-cyclocarbosiloxane was synthesized, and at the second stage, dichlorobicyclocarbosiloxane was pro-duced by HFC reaction of tetrachlorocyclocarbosiloxane with dihydroxydiphenylsilane at 1:1 ratio of initial components, all other conditions being the same. In both cases, the reaction proceeds in accor-dance with the general scheme as follows:

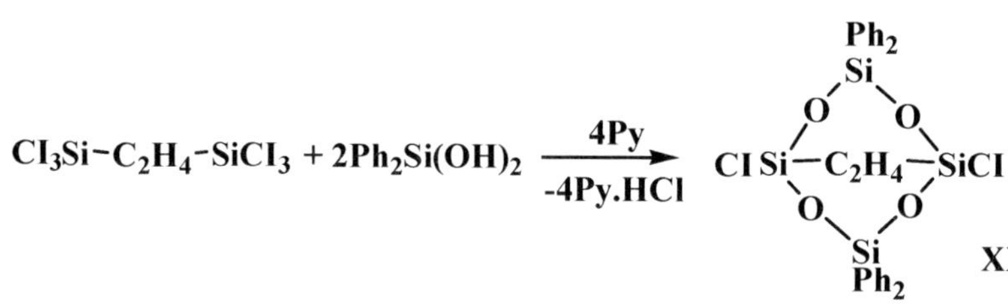

XXII

Scheme 25.

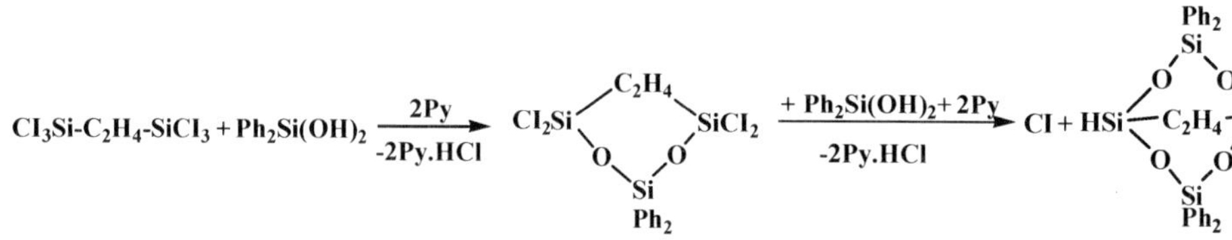

XXII

Scheme 26.

Composition and structure of synthesized dichlorobicyclocarbosiloxane (compound 1, Table 2) of stru-cture XXII were determined from IR and ^{1}H NMR spectra that displayed a singlet reflex with chemical shift at $\delta = 1.078$ ppm, typical of methylene groups, due to the molecule symmetry.

Hydrolysis of dichlorobicyclocarbosiloxane in the neutral medium synthesized corresponded dihydro-xybicyclocarbosiloxane by the scheme as follows:

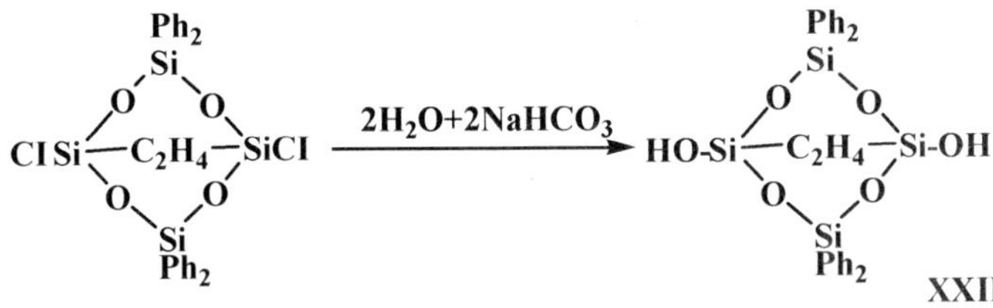

XXII

Scheme 27.

Some physical and chemical parameters of synthesized bicyclocarbosiloxanes are shown in Table 2.

Table 2. Physical and chemical parameters of difunctional organobicyclosiloxanes

№	Compound	Yield, %	T_{boil}, ^{0}C (P, mm.Hg.)	T_{melt}, ^{0}C	^{1}H chemical shift	References
1	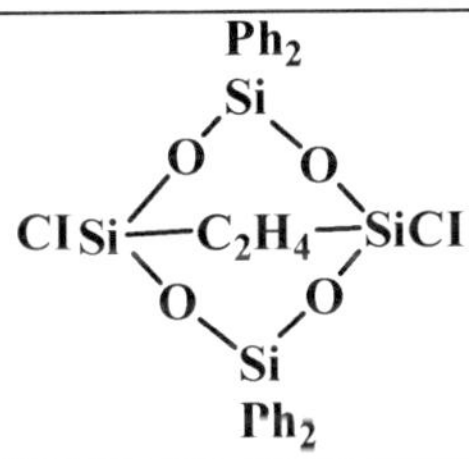	44 (69)	240-250 (0,02)	172-179	1,078	[22,23]
2	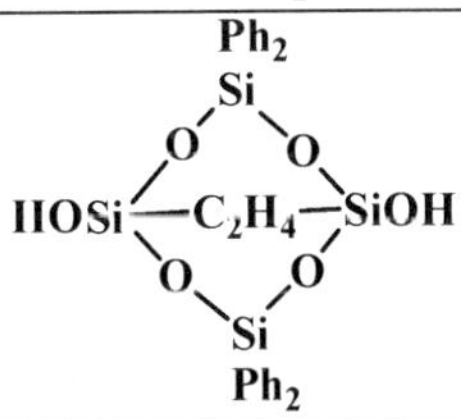	65	–	218-219	–	[22,23]

The Synthesized heterobi(tri- and tetra-)cyclic organosilicon compounds with functional groups are interesting products for production of cyclolinear copolymers.

References

[1] Andrianov K.A., Jenchelskaya S.M., and Petrashko Yu.K., *Zh. Obshch. Khim.*, 1958, vol. 28(3), p. 685. (Rus)

[2] Sokolov N.N., Methods of Poly(organosiloxane) Synthesis, Moscow –Leningrad, *Gosnergoizdat,* 1959, p. 103. (Rus)

[3] Andrianov K.A., Polymers With Inorganic Backbones of Macromolecules, Moscow, Izd. AN SSSR, 1962, p. 144. (Rus)

[4] Andrianov K.A., Jenchelskaya S.M., and Petrashko Yu.K., *Plast. Massy,1960*, No. 3, p. 20. (Rus)

[5] Meladze S.M., Mukbaniani O.V., Makarova N.N., and Khananashvili L.M., *Zh. Obshch. Khim*., 1980, vol. 50(11), p. 2493. (Rus)

[6] Mukbaniani O.V., Khananashvili L.M., and Inaridze I.A., *Int. J. Polym. Mater*. 1994, vol. 24, p. 211.

[7] Andrianov K.A., Nogaideli A.I., Makarova N.N., Raspopova K.N., and Mukbaniani K.A., *Doklady AN SSSR*, 1975, vol. 224(4), p. 825. (Rus)

[8] Mukbaniani O.V., *Candidate's Dissertation on Chemistry*, Tbilisi State University, 1976, Tbilisi, GSSR. (Rus)

[9] Andrianov K.A., Makarova N.N., and Chizhova N.V., *Izv. AN SSSR, Ser. Khim*., 1975, p. 2554. (Rus)

[10] Mukbaniani O.V., *Doctor's Dissertation on Chemistry*, Tbilisi State University, 1993, Tbilisi, Georgia.

[11] Andrianov K.A., Nogaideli A.I., Makarova N.N., and Mukbaniani O.V., *Soobshch. AN GSSR*, 1976, vol. 81(2), p. 349. (Rus)

[12] Mukbaniani O.V., Khananashvili L.M., Karchkhadze M.G., Tkeshelashvili R.Sh., and Mukbaniani N.O. In: *Synthesis and Properties of Polymers*, Nova Science Publishers, Inc., Commak., 1996, p. 89.

[13] Klementiev I.Yu., *Candidate's Dissertation on Chemistry*, M.V. Lomonosov Institute of Fine Chemical Technology, 1979, Moscow, USSR. (Rus)

[14] Andrianov K.A., Zhdanov A.A., Kurasheva N.A., and Khyinku E.S., *Doklady AN SSSR*, 1977, vol. 236(2), p. 354. (Rus)

[15] *Patent No. 3,378,575* (USA).

[16] *Patent No. 3,222,369* (USA).

[17] *Patent No. 3,317,578* (USA).

[18] Andrianov K.A., Zhdanov A.A., Kurasheva N.A., and Khyinku E.S., *Doklady AN SSSR*, 1978, vol. 240(3), p. 591. (Rus)

[19] Zhdanov A.A., Khyinku E.S., Kurasheva N.A., and Shkvoler V.E., *Doklady AN SSSR*, 1980, vol. 250(6), p. 1392. (Rus)

[20] Mukbaniani O.V., Koyava N.A., Karchkhadze M.G., Shengelia M.G., and Khananashvili L.M., *Int. J. Appl. Polym. Mater*., 2001, vol. 82, p. 3142.

[21] Mukbaniani O.V., Koyava N.A., Karchkhadze M.G., Shengelia M.G, and Khananashvili L.M., In: *Radical and Ion Reactions. Problems and Ways of Their Solution*, Ed. by G.E. Zaikov, 2001, Nova Science Publishers, Inc, p. 99.

[22] Inaridze I.A., Doctor's Dissertation on Chemistry, Tbilisi State University, 1993, Tbilisi, Georgia.

[23] Mukbaniani O.V., Inaridze I.A., and Meladze S.M., *Thes. Rep. IV Republ. Research Conference "Chemistry and Scientific Progress"*, 1988, Kutaisi, p. 64. (Rus)

In: New Topics in Monomer and Polymer Research
Editors: A. D'Amore and G. Zaikov, pp. 135-151
ISBN: 1-60021-436-3

Chapter 9

Hydrosilylation and Dehydrocondensation Reactions of Methylhydridesiloxane to Acrylic and Methacrylic Acids

O. Mukbaniani[1,2], G. Zaikov[3], N. Pirckheliani[2], T. Tatrishvili[1], S. Meladze[1], Z. Pachulia[2], M. Labartkava[1]

[1]I. Javakhishvili Tbilisi State University,
I. Chavchavadze Ave., 1, 0128 Tbilisi, Georgia
[2]Sukhumi State University, Branch of Tbilisi State University,
Djikia Str., 12, 0187; Tbilisi, Georgia
3Institute of Biochemical Physics Russian Academy of Sciences
4 Kosygin Street, Moscow 119991, Russia

Abstract

The reaction of hydrosilylation of α,ω–bis(trimethylsiloxy)methylhydridesiloxane (n≈53) with acrylic acid at various ratio of initial compounds, in the presence of Platinum hydrochloric acid has been investigated. It was established that in the presence of Platinum hydrochloric acid competitive dehydrocondensation reactions take place with formation of oligomers with various link of chain. In high stages of reaction, three-dimensional systems were obtained. During dehydrocondensation the reaction order, reaction rate constants and activation energy were found.

The structures of synthesized oligomers were determined by IR and NMR spectra data. Gel perme-ation chromatographic, differential scanning calorimetric and X-ray analysis of synthesized oligo-mers were carried out.

Key words: dehydrocondensation, hydrosilylation, methylhydridesiloxane.

INTRODUCTION

In literature there are many information about hydrosilylation of acrylic acid [1-4]. By A.D. Petrov the reaction of hydride addition of triethylsilane to acrylic acid in the presence of Spier catalyst has been investigated and it was shown that the reaction proceed in "1-4" position, with formation intermediate product, by rearrangement of this product trimethylsilylated ether of propionic acid with 56% yield was obtained. The reaction proceeds according to the following scheme [1]:

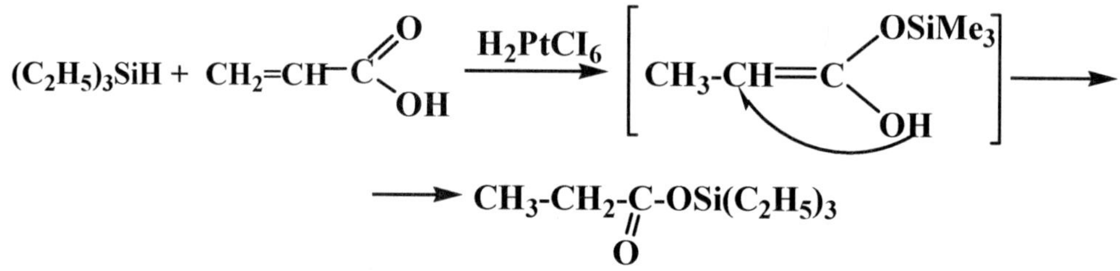

By the same author it was shown that the hydrosilylation acrylic methyl ether with triethylsilane proceeds in "1-4" position with formation unsaturated silylated methyl ether [2].

The reaction of hydride addition of α,ω–dihydridedimethylsiloxanes with acrylic and methacrylic acid has been investigated by K.A. Andrianov and his co-workers, in the presence of Platinum hydrochloric acid and it was shown that hydrosilylation proceeds on an unsaturated C=C bond with Farmer rule receiving corresponding dicarboxylic acids. The reaction proceeds according to the following scheme [3,4]:

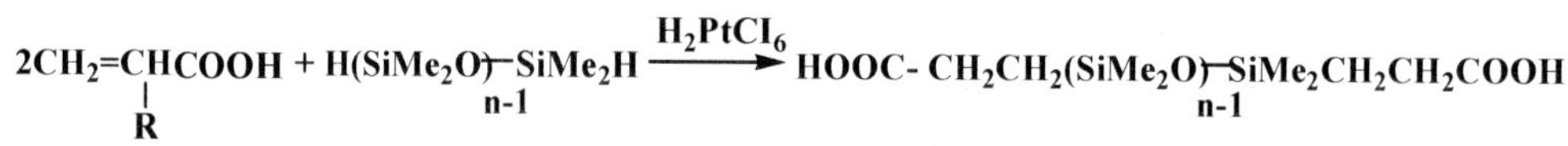

Where: R=H, Me; n=2÷11.

The structure of synthesized silicon containing dicarbon acids was determined by means of finding of molecular masses and IR spectra data.

The reaction of hydrosilylation of acrylic and methacrylic acid with 1,4-bis(dimethylsilyl)benzene in the presence of hydroquinone and Spier catalyst or platinum on the carbon was studied and it was shown that the reaction proceeds mainly by dehydrocondensation way, with formation of corres-ponding acyloxysilanes, according to the following scheme [5]:

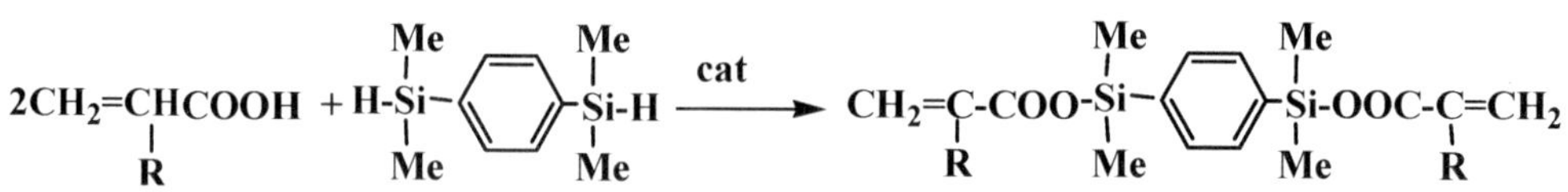

Where: R=H, Me.

In hydrosilylation reactions the hydroquinone was used for preventing the reactions of poly-merization process.

The reaction of hydrosilylation of methylhydridesiloxanes to unsaturated vinyl derivatives (alkenes, acrylic acid, acrylic ethers) in the presence of Spier catalyst has been investigated by the authors [6] and methylsiloxane oligomers with alkyl, carboxyl and ester side-groups has been obtained, accor-ding to the following scheme:

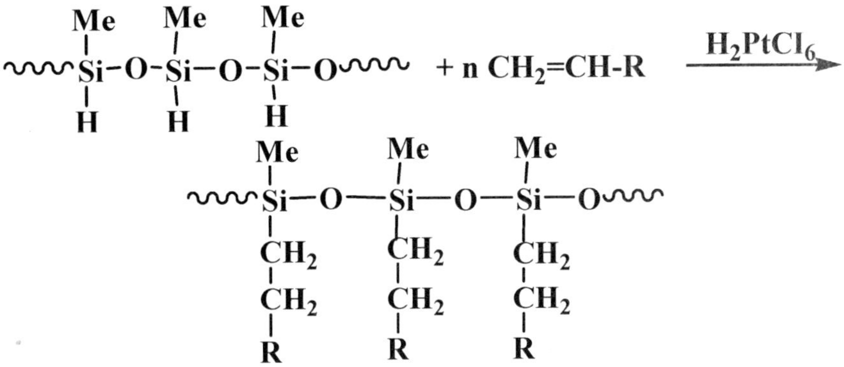

Where: $R=C_nH_{2n+1}$, -COOH, -COOAlk.

It is also known, that the reaction of catalytic dehydrocondensation between ≡Si-H containing compounds and hydroxyl containing organic compounds proceeds with isolation of hydrogen [7-9]. In the reactions of catalytic dehydrocondensation of hydrideorganosilanes with hydroxyl containing organic compounds are used such catalysts as colloid nickel, anhydrous zinc chloride, Platinum hydrochloric acid [10,11]. It was shown [9] that in the reaction of dehydrocondensation of dihydrideorganocyclosiloxanes with dihydroxyorganocyclosiloxanes Platinum hydrochloric acid works as caustic potassium and activation energy have the same values. On the other hand, it is known that Platinum hydrochloric acid is catalyst of hydrosilylation [12]. The reaction capacity of ≡Si-H bond determined by catalyst nature and amount, reaction temperature, reactant concentration, solvent character, etc [7, 8]. By the authors [9] it was shown that during dehydrocondensation of hydride-containing organosiloxanes with hydroxyl-containing organosilanes and siloxanes in the presence of powder-like caustic potassium as a nucleophilic catalyst the competing reaction of dialkylsiloxane elimination docs not take place that is explained by a high rate of the reaction of dehydrocondensation process.

Experimental Part

The starting materials for the synthesis of comb-type methylsiloxanes oligomers were used α,ω– bis(trimethylsiloxy)methylhydridesiloxane, acrylic and methacrylic acids.

The initial α,ω–bis(trimethylsiloxy)methylhydridesiloxane with the degree of polymerization n≈35, acrylic and methacrylic acids was received from Fluka. Before reaction, they acrylic and methacry-lic acids were distilled. The organic solvents used in reaction hydrosilylation were cleaned by dry-ing and distillation.

IR spectra all of samples have been taken on an UR-20 instrument from KBr pellets, while the ^{1}H NMR spectra on a "Perkin-Elmer" at operating frequency of 250 MHz. All spectra were obtained with the use of $CDCl_3$ as solvent and internal standard. A Perkin–Elmer

DSC-2 differential scanning calorimeter was used to determine TGA and the thermal transitions (T_g) were read at the maximum of the endothermic or exothermic peaks. Heating and cooling scanning rates were 10^0C/min.

Gel-permeation chromatography investigation was carried out with the use of Waters Model 6000A chromatograph with an R 401 differential refractometer detector. The column set comprised 10^3 and 10^4 Å Ultrastyragel columns. Sample concentration was approximately 3% by weight in toluene and typical injection volume for the siloxane was 5 μL. standardization of the GPC was accompli-shed by the use of styrene or polydimethylsiloxane standards with the known molecular weight.

Wide-angle X-ray diffractograms were taken on a "DRON-2" instrument. A-CuK_α was measured without a filter, the angular velocity of the motor $\omega \approx 2$ deg/min.

Dehydrocondensation Reaction of α,ω–Bis(Trimethylsiloxy)Methylhydridesiloxane with Acrylic Acid

Catalytic dehydrocondensation reaction was carried out in a two-necked flask equipped with a tube for catalyst inlet, reflux condenser connected with a gasometer. Between the gasometer and re-flux condenser a cold trap and wash bottle were installed.

The reaction products were placed into the flask and thermostated in an oil bath until constant temperature was set up. Then the catalyst 0,1 M solution of Platinum hydrochloric acid in tetra-hydrofuran ($5 \div 9 x 10^{-5}$ g per 1,0 g of starting substance) was introduced. After that, hydrogen started release; hydrogen waªËcollected in the gasometer. After the reaction has been finished 5 ml anhyd-rous toluene was added, was filtered and the solvent was eliminated.

Hydride Addition of α,ω–Bis(Trimethylsiloxy)Methylhydridesiloxane to Acrylic Acid

The hydrosilylation reaction was carried out in a three-necked flask equipped with stirrer, reflux condenser and calcium chloride drying tube. The initial reagents were placed in the flask and ther-mostated in an oil bath until constant temperature was achieved. 0,1 M solution of platinum chlo-rohydric acid in tetrahydrofuran ($5 \div 9 x 10^{-5}$ g per 1,0 g of starting substance) was used as a catalyst. The reaction carried out in an argon atmosphere in anhydrous toluene solution, at temperature range 60^0C. Then 5 ml anhydrous toluene to the reaction mixture was added, filtered connected to the vacuum and the toluene and unreacted acrylic acid was removed at $30 \div 40^0$C.

The hydrosilylation reaction in the presence of platinum on the carbon carried out in the same manner.

Results and Discussion

As it is seen from the literary review, the reactions of hydride addition ≡Si-H bond containing com-pounds to acrylic and metacrylic acid proceeds on different directions and disagrees to each other. The presented paper deals with synthesis and studies of the properties of comb-type organosi-loxanes containing carboxyl groups in the side chain.

For the purpose of synthesis of organosiloxanes oligomers with carboxyl groups in the side chain the reaction of hydride addition of α,ω-bis(trimethylsiloxy)methylhydridesiloxanes (n≈35) to ac-rylic and methacrylic acids at 1:35 ratio of initial compounds, in the presence of catalysts Platinum hydrochloric acid or Pt/C (5%), at various temperature as in dry toluene solution (50-60%) so in melt condition has been investigated.

Preliminary heating of the initial compounds for 4 h in the presence of 0,1 M of platinum chlo-rohydric acid solution in tetrahydrofuran ($\sim 5 \div 9 \times 10^{-5}$ g per gram of reaction substances) in the tem-perature range 50-70^{0}C showed that polymerisation of acrylic or methacrylic acids, breaking of siloxane backbone, elimination of methane or hydrogen in methylhydridesiloxanes or other changes do not take place. Besides there are no changes in the NMR spectra of initial compounds. As it is known from literature [13] Platinum hydrochloric acid except of the reactions of hydrosilylation catalysed the reaction of dehydrocondensation. Because of that, for completely characterization of the process during the reaction the conversion of hydrogen was observed.

The reaction of hydrosilylation investigated without solvent in the presence of catalysts at 70^{0}C temperature and it was shown, that at low stages of hydrogen conversion ~10÷20% the reaction proceeds with formation of three-dimensional systems. In this case, the depth of hydrosilylation reaction is about ~60%. For preventing the cross-linking processes, the reactions were conducted in ambience of dry toluene solution.

During the reaction in solution, it was established, that at ~35% hydrogen conversions (Figure 1) the cross-linking processes take place. By determination of active ≡Si-H groups from soluble part of reaction products it was established, that the depth of hydrosilylation is about ~50%. This point that in the presence of Platinum hydrochloric acid in dilute solution of toluene, besides of hydrosily lation concurrent runs the reaction of dehydrocondensation ~30% (Figure 1, curve 1).

For the purpose of synthesis oligomers, well soluble in organic solvents the same reaction was studied in the presence of platinum on the carbon (5%) in dry toluene solution, but it was shown that really the reaction of dehydrocondensation proceeds less activity 9-10%, Figure 1 (curve 2), but the three dimensional systems are obtained. Hydrosilylation in this condition proceeds about ~54%. Therefore, in both cases the reaction proceeds by two concurrent ways, with formation various- linked oligomers, according to the following scheme:

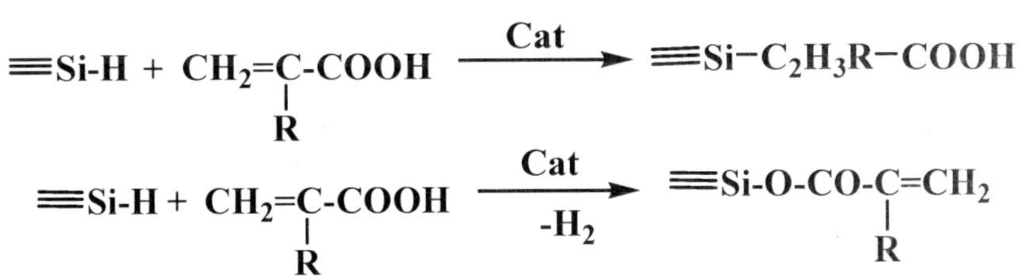

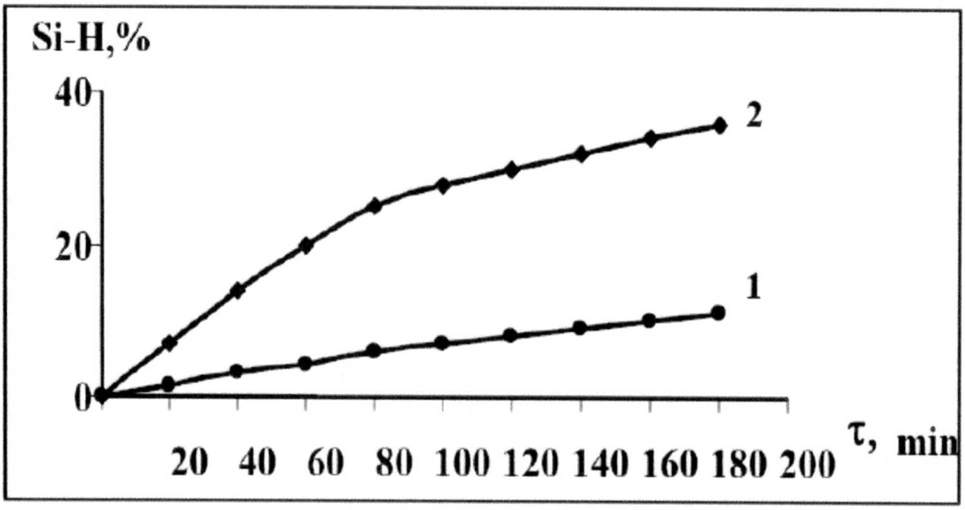

Figure 1. Dependence of changes of active ≡Si-H% groups on the time, upon of dehy-drocondensation reaction of α,ω–bis(trimethylsiloxy)methylhydridesiloxane to acrylic acid at 60^0C temperature. Where curve 1 corresponds to catalyst Pt/C, curve 2 corres-ponds to - H_2PtCl_6.

With an increase of dilution of reaction solution the hydrosilylation and dehydrocondensation reaction rates decreases, but increases the soluble part of reaction products. Generally the reaction proceeds according to the following scheme:

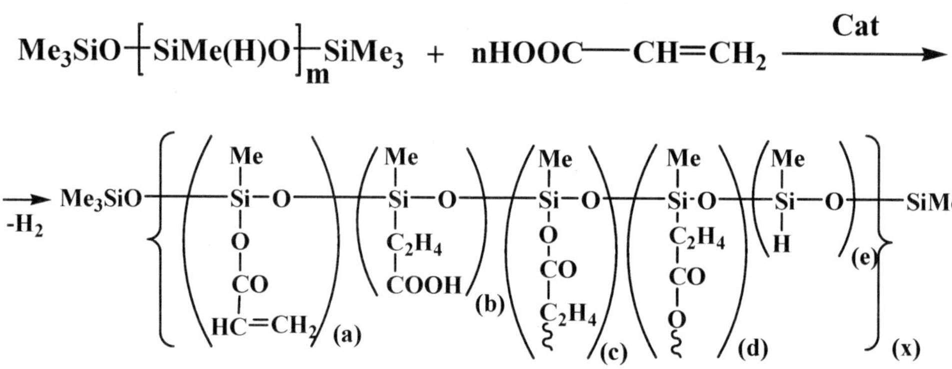

Where: m≈35; [(a)+(b)+(c)+(d)+(e)]x(x)=m. Cat – H_2PtCl_6, I (ratio 1:35); II (ratio 1:17,5) Pt/C – I^1 (ratio 1:35).

From cross-linked polymers the soluble part was isolated with the yield ~50-66%. They are colourless transparent viscous products soluble in ordinary organic aromatic type solvents. Some physical-chemical properties and the yields of synthesized branched oligomers are presented in Table 1.

It must denote that for a time the synthesized oligomers undergo to self-catalysed structuring process. From oligomer I after 24 hours was isolated soluble part III with the yield ~13%.

Table 1. Some physical and chemical properties and the yields of branched oligomers

Oligomer, №	Yield, %	Ratio	η_{sp}^{*}	H_2 Conv., %	Depth of hydrosilylation, %	d_1, Å	$\bar{M}_\omega x10^{-3}$	T_g, ^{0}C
I	57	1:35	0,08	30	50	8,85	7,2	-67
II	49	1:17,5	0,06	23	49	-	-	-
I^1	66	1:35	0,08	10	54	8,23	-	-70
III	13	-	0,07	-	-	8,92	60,8	-23

*1% solution in Toluene at 25^{0}C.

For oligomer III, ^{1}H and ^{13}C NMR investigation carried out. As it is seen from Figure 2 one can observe triplet signal, with chemical shift δ≈1,25 ppm (J=6,0 Hz) characteristic for hydrogen protons in ≡Si–CH_2- group. One can see also triplet signal for hydrogen protons in -CH_2-CO- group with chemical shift δ≈2,5 ppm and singlet signals with chemical shift δ≈0,1 and δ≈0,3 ppm for methyl protons in -$SiMe_3$ and ≡SiMe groups accordingly. In spectra one can see complicated mul-tiplet characteristic for vinyl groups with chemical shift δ≈5,80÷6,1 ppm, which indicate that the reaction proceeds partially by dehydrocondensation way. There is low intensity singlet signal with δ≈4,2 ppm characteristic for unreacted ≡Si-H bond. One can see also, characteristic signal for residual acids protons with chemical shift δ≈10,2 ppm and multiplet signals of phenyl protons in toluene with chemical shift δ≈7,2÷7,6 ppm, which indicate the traces of toluene in synthesized oligomers.

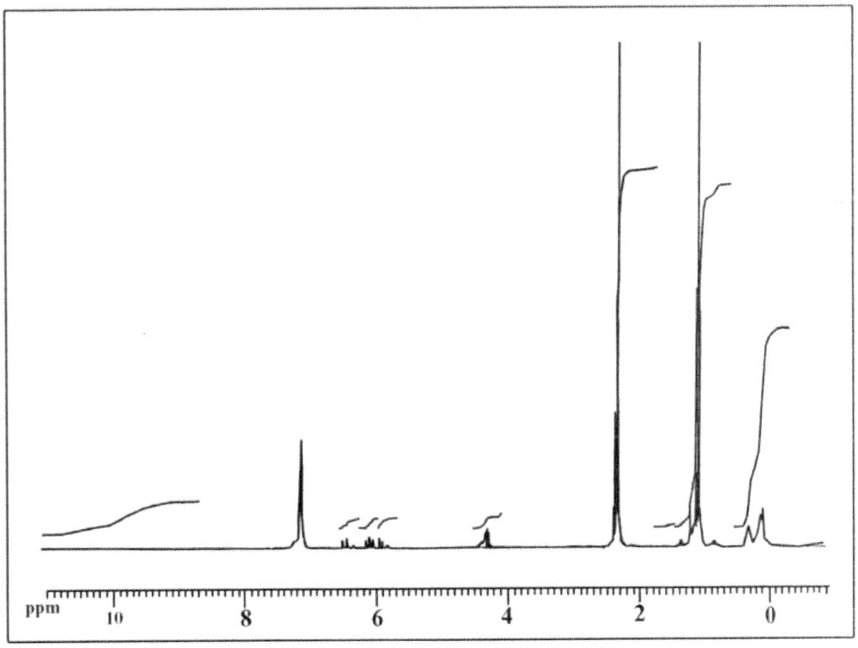

Figure 2. ^{1}H NMR spectra of oligomer III (solvent and standard $CDCl_3$).

In ^{13}C spectra of oligomer III (Figure III) it's well seen the signal for carbon in residual acid car-bonyl group with chemical shift δ≈179,3 ppm; signal with chemical shift δ≈7,5 ppm in ≡SiMe group, signal with chemical shift δ≈6,9 ppm in -$SiMe_3$; signal with chemical shift δ≈25,9 ppm characteristic for carbon in ≡Si–CH_2- group; signal with chemical shift δ≈27,5 ppm characteristic for carbon in methylene group in -CH_2-COOH fragment. Concerning of the reaction on the other direction, by dehydrocondensation way, is proved by presence in ^{13}C spectrum the signal with chemical shift δ≈171,7 ppm characteristic for carbon in free from proton carboxyl group ≡Si–O-CO-. Obtaining of cross-linked systems may be explained by intermolecular hydrosilylation and dehydrocondensation reactions:

$$-C_2H_4-COOH + H-Si\equiv \xrightarrow[-H_2]{Cat} \equiv Si-C_2H_4-CO-O-Si\equiv$$

$$\equiv Si-O-CO-CH=CH_2 + H-Si\equiv \xrightarrow{Cat} \equiv Si-O-CO-C_2H_4-Si\equiv$$

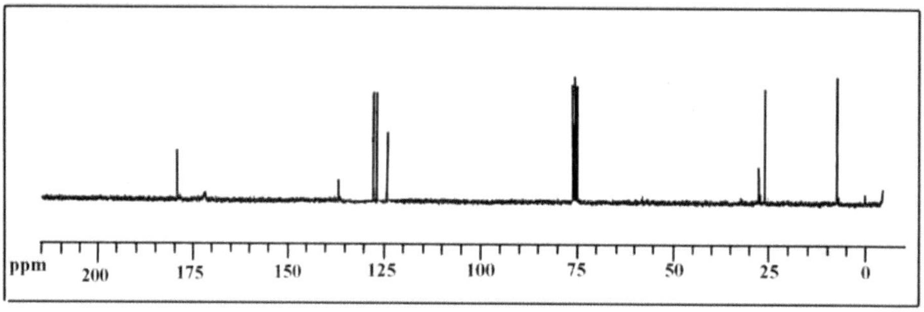

Figure 3. ^{13}C NMR spectra of oligomer III (solvent and standard $CDCl_3$).

In the IR spectra of oligomers one can see the absorption bands characteristic for ≡Si-O-Si≡ and ≡Si-O-C≡ bonds at 1020 and 1080 cm^{-1}. In the IR spectra one can see the absorption bands for ≡SiMe, for unreacted ≡Si-H bonds and intensive signal for carbonyl ethereal groups, signals for double bonds at 1275, 2165, 1720 and 1640 cm^{-1} accordingly.

It must be denote, that in IR spectra there is no absorption bands characteristic for associated and no associated hydroxyl groups. From which it is evident, that hydroxyl group take part in intermo-lecular dehydrocondensation as it is stated above, which led to formation of three dimensional or cross-linked systems and in intramolecular cyclization according to the following scheme:

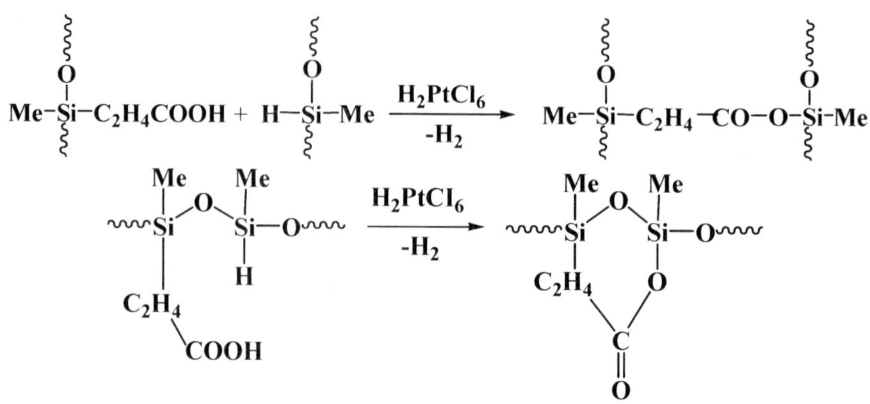

Where: $-C_2H_4- = -(CH_2)_2-$.

On the other side if the reaction proceeds by dehydrocondensation way, in this case in synthesized oligomers resonance signals for unsaturated vinyl fragments, with chemical shift in the range 5,5÷6,2 ppm, must be observed. But in the spectra Figure 2 and 3 there are low intensity resonance signals, which show that the fragment **≡Si-O-CO-CH=CH$_2$** may be entered in intra- and inter-molecular hydride addition by the scheme:

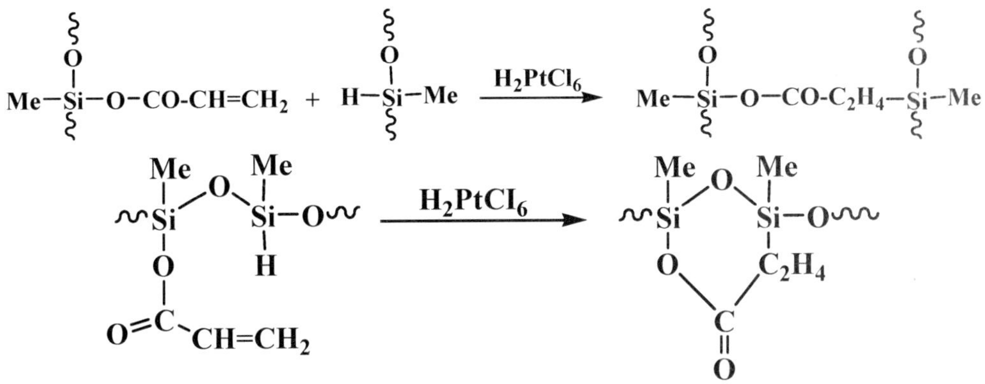

Where: $-C_2H_4- = -(CH_2)_2-$.

For oligomer, I and III wide angle X-ray investigation has been carried out. It was shown that oligomers are represented as one phase amorphous products, where the value of interchain distances changes in the range $d_1 \approx 8{,}23\text{-}8{,}92$ Å.

The synthesized oligomers were studied by GPC method. As it is seen from Figure 4 the GPC cur-ves of soluble part of oligomers I and III sharply distinguished from each other. Molecular weight distribution of oligomer I has bimodal character (Figure 4, curve 1) and for oligomer III polymodal character (figure 4, curve 2), from which it's evident that during hydrosilylation reaction dehydro-condensation reactions takes place, that is in a contradiction with earlier published data [3,4]. Average weight molecular masses of oligomer I is equal to $\overline{M}_\omega \approx 7{,}2\text{x}10^3$ and polydispersity D≈9,4. For irregular branched oligomer III average weight molecular masses of oligomer III is equal to $M_\omega \approx 60{,}8\text{x}10^3$ and polydispersity D≈35,3.

Analogically hydrosilylation reaction of α,ω–bis(trimethylsiloxy)methylhydridesiloxane to meta-crylic acid in the presence of catalyst – Platinum hydrochloric acid H_2PtCl_6 (0,1 M solution in tetrahydrofuran) at 1:35 ratio of initial compounds, at various temperature, both in a melt and in a dry toluene solution (50-60%) has been investigated.

It was shown that at 40^0C in case of 2% conversion of hydrogen the cross-linking processes take place. From soluble part of cross-linked products by determination of amount of active ≡Si-H group it was established, that the depth of hydrosilylation is about ~84%. With an increase of temperature from 50^0C and to 60^0C the part of the reaction of dehydrocondensation increases from ~3,8% up to ~15,0% (see Figure 5). The part of hydrosilylation reaction at 60^0C is equal to ~82%. So, during the reaction of α,ω–bis(trimethylsiloxy)methylhydridesiloxane to methacrylic acid in the presence of catalyst - H_2PtCl_6, besides of hydrosilylation reaction (both direction – Farmer rule and Markov-nikov rule), catalytic dehydrocondensation takes place.

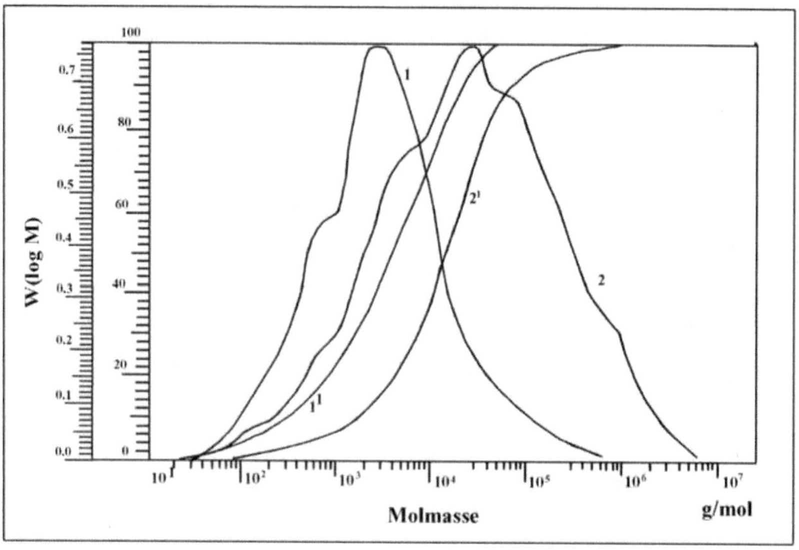

Figure 4. Gel permeation chromatographic curves of oligomers. Where curve; 1 corresponds to oligomer I and curve 2 corresponds to oligomer III.

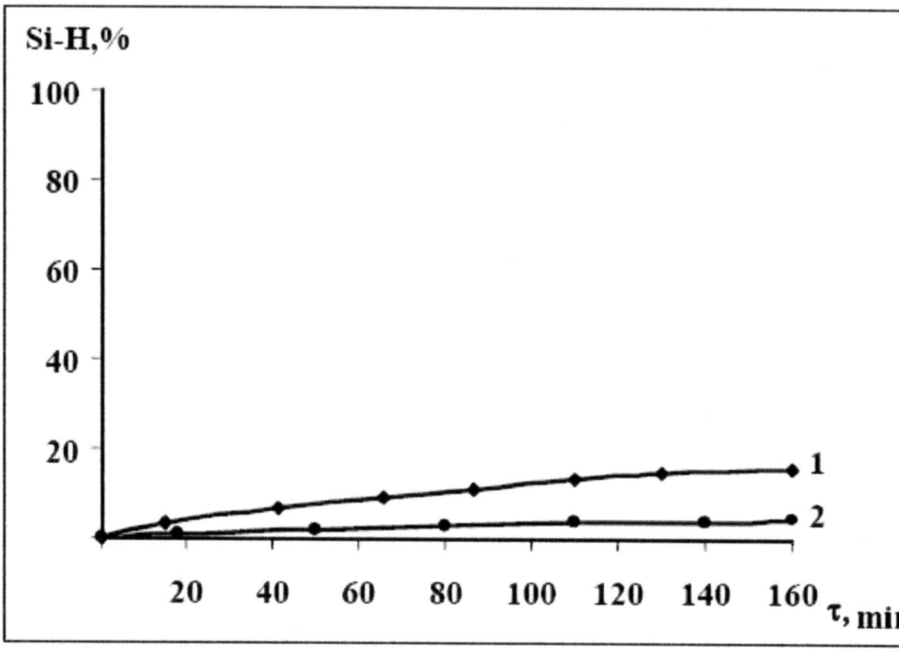

Figure 5. Dependence of changes of concentration active ≡Si-H% groups on the time, upon of dehydrocondensation reaction of α,ω–bis(trimethylsiloxy)methylhydridesiloxa-ne to methacrylic acid at 60^0C temperature. Where curve 1 corresponds to 50^0C, curve 2 60^0C (in the presence of H_2PtCl_6).

During the dilution the depth of dehydrocondensation and hydrosilylation reactions decreases and the yield of soluble part rise. The reaction of methylhydridesiloxane with methacrylic acid proceeds with formation of various-linked products, according to the following scheme:

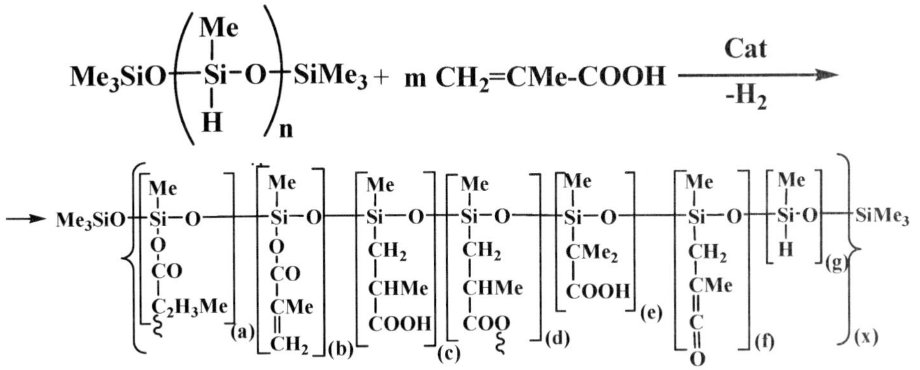

Where: [(a)+(b)+(c)+(d)+(e)+(f)+(g)](x)=n; 40^0C - IV^1; 50^0C - IV^2 and 60^0C - IV^3.

From cross-linked polymers, obtained branched oligomers are viscous products well soluble in aromatic solvents, with $\eta_{sp}\approx 0{,}04$-$0{,}06$. Some physical-chemical properties and the yield of oligo-mers IV^1-IV^3 are presented in Table 2.

Table 2. Some physical and chemical properties and the yield of oligomers IV^1-IV^3

#	Yield, %	Ratio	$\eta^*_{sp.}$	Reaction t-re T^0C	Conversion of H_2, %	Hydrosilylation, %	d_1,Å
IV^1	68	1:35	0,04	40	2,0	84	-
IV^2	66	1:35	0,05	50	3,8	88	-
IV^3	64	1:35	0,06	60	15	82	9,07

*In 1% solution of toluene at 25^0C.

For synthesised oligomers where studied by ^1H and ^{13}C NMR spectral data. Based on spectral data of reaction products proposed, that in generally hydride addition of methylhydridesiloxane to acrylic and methacrylic acid may be proceeds by several ways, see scheme 1:

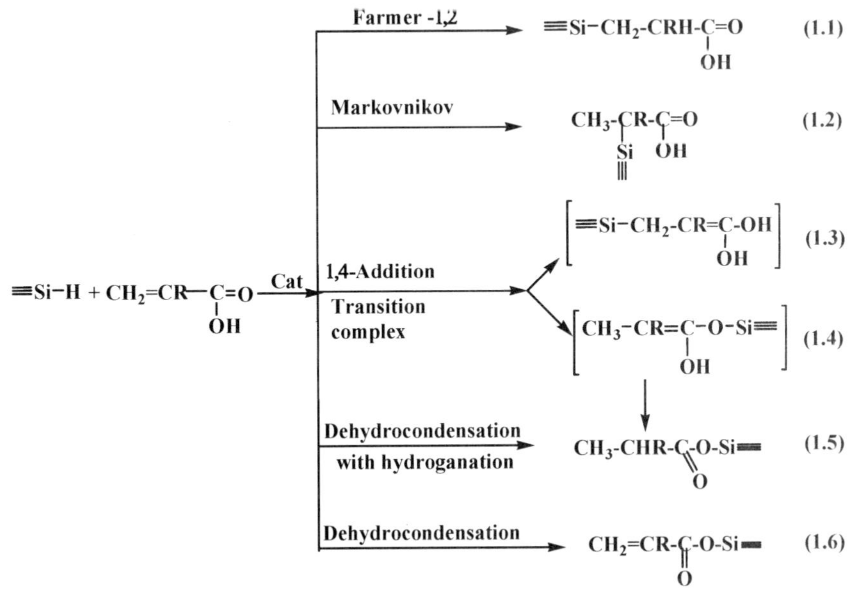

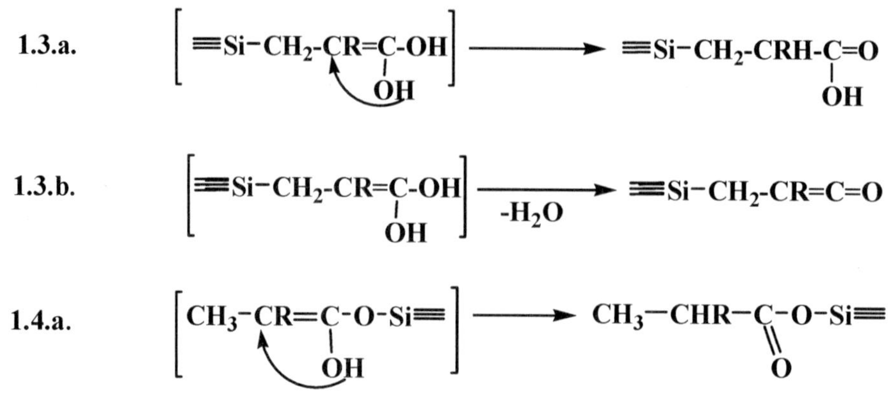

Scheme 1. The possible scheme of hydrosilylation of methylhydridesiloxane to acrylic and methacrylic acids.

In ^{1}H NMR spectra of oligomer IV^3 one can observe singlet signal with chemical shift δ≈1,21 ppm for methyl protons in =CH-Me group, broadened singlet signal for protons of methyl group in C-Me group, signal with chemical shift δ≈1,22 ppm in α-addition product (Markovnikov rule, scheme 1.2), signal with chemical shift δ≈0,8 ppm corresponds to protons of methylene group, β-addition product in ≡Si-CH_2- fragment, which creation may be examine as structure obtained by regrouping of intermediate product (transition complex - 1,4-addition according to the Eltekov rule (scheme 1.3a).

The structure of the product obtained by above-mentioned scheme 1.3a is similar to the structure of the product obtained by Farmer rule (1,2-addition). Chemical shifts and multiplicity of remaining protons are: for methine group (≡CH) sestet (in case R=Me) with δ≈2,6 ppm (J=6,83 Hz), for methyl group singlet at δ≈1,21 ppm, (J=7,07 Hz) and for protons in carboxyl group broadened signal with chemical shift at δ≈10-11 ppm. In case of another course of reaction, after which elimination of molecule of water under the schema 1.3 takes place, with formation ketene group, their chemical shifts are following: for protons in methylene ≡Si-CH_2- group singlet signal with chemical shift δ≈0,78 ppm, for methyl protons in –CMe=C=O fragment singlet signal with chemi-cal shift δ≈1,96 ppm. It must be denote, that the authors [1] suggested direction of the reaction with obtaining of transition complex according to the scheme 1.4 earlier. Obtained transition complex then undergoes regrouping according to the scheme 1.4a with obtaining simultaneous dehydrocondensation-hydrogenation product like to the scheme 1.5. However, processing the reac-tion to above-mentioned scheme is improbable, because released hydrogen during dehydroconden-sation reaction, is eliminated away from reaction mixture.

Obtaining three various hydrosilylation products with methacrylic acid proved by ^{13}C NMR spectra data. Each of adduct are characterized with following chemical shifts: for hydrosilylation product by Markovnikov rule (scheme 1.2) resonance signal for methyl protons at δ≈18,11 ppm, signal for quaternary carbon nucleus at δ≈34,11 ppm and signal for carbon nucleus in carboxyl group at δ≈172,74 ppm. In hydrosilylation product by Farmer rule (1,2-addition, scheme 1.1), one can obser-ve signal for carbon nucleus with chemical shift at δ≈18,98 ppm in ≡Si-CH_2- fragment, signal at δ≈29,58 ppm for carbon nucleus in =CH- group and signal for methyl group at δ≈22,92 ppm in Me-CH- fragment.

One can observe analogically value of signals for compound containing ketene groups. Signal with chemical shift δ≈18,98 ppm characteristic for carbon nucleus of methylene group (≡Si-CH_2-), signal at δ≈30,2 ppm for carbon in methyl group [-C(Me)=C=O], signal for tertiary carbon nucleus at δ≈ 135,6 ppm [-C(Me)=C=O] and signal at δ≈183,43 ppm for carbon nucleus in the group =C=O.

As it is seen from ^{1}H and ^{13}C NMR spectra data hydrosilylation reaction of methylhydridesiloxanes to acrylic and methacrylic acid distinguished from each other and from literature data [1]. With acrylic acid hydrosilylation mainly proceeds by dehydrocondensation and hydrosilylation (Farmer rule (scheme 1.1 and 1.3a). In contrast of this, hydrosilylation to methacrylic acid proceeds almost in all direction presented on the scheme 1.

For fully characterization of hydride addition of methylhydridesiloxane to acrylic and methacrylic acid by quantum-chemical half empiric AM1 method [14], for all initial, intermediate and final products, in model reaction of hydrosilylation of methyldimethoxysilane to acrylic and methacrylic acid, it was calculated heats of formations (ΔH_f) and change of enthalpy of model reactions ($\Delta\Delta H$) proceeding with various concurrent scheme 2:

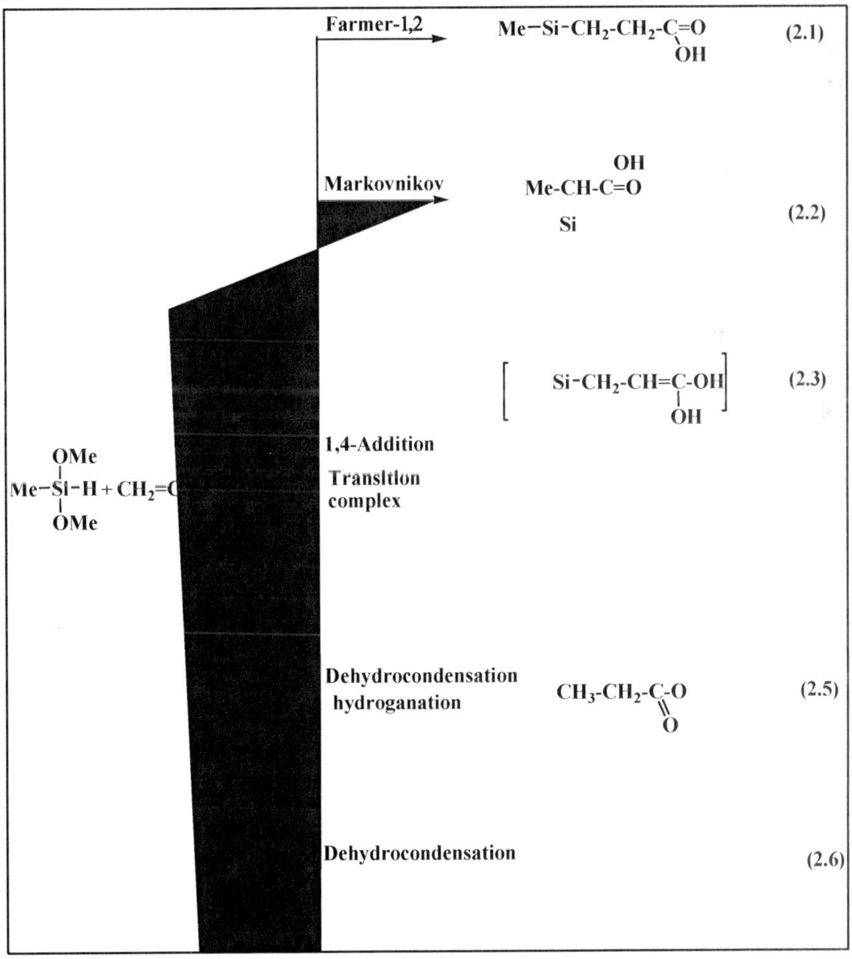

Scheme 2. Continued.

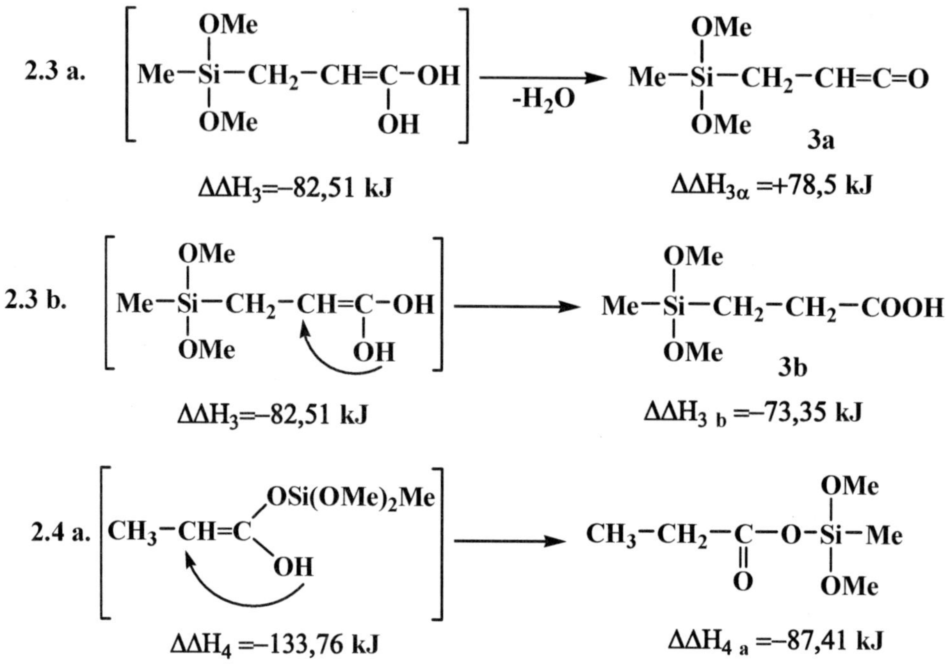

Scheme 2. Change of enthalpy of model reactions of hydrosilylation of methyldimethoxysilane to acrylic acid.

As it is seen from the scheme 2, starting from calorific effects of all possible direction of hydro-silylation reaction of methyldimethoxysilane to acrylic acid thermodynamically is more profitable course of the reaction according to the Farmer rule (1,2-addition). It must be denoted that the pro-duct obtained by regrouping of intermediate product (1,4-addition, scheme 2.3) similar to Eltekov rule (scheme 2.3b), identical to product obtained by Farmer rule (scheme 2.1) and they have the same value of calorific effects [$\Delta\Delta H_1=\Delta\Delta H_3+\Delta\Delta H_{3a}=$-82,51 kJ+(-73,35 kJ)=-135,85 kJ]. Obtaining of ketene by regrouping of intermediate product according to the scheme 2.3a is most probable all.

As it is seen from ^{1}H and ^{13}C NMR spectral data of oligomer III, in the reaction of hydrosilylation of methylhydridesiloxanes to acrylic acids obtaining of ketenes is not observed.

The structure of the product obtained by dehydrocondensation-hydrogenation reactions (scheme 2.5) is identical to the product obtained by regrouping of intermediate product (scheme 2.4) accor-ding to the Eltekov rule. Possible scheme of the reaction in enlarged aspect may be represented as:

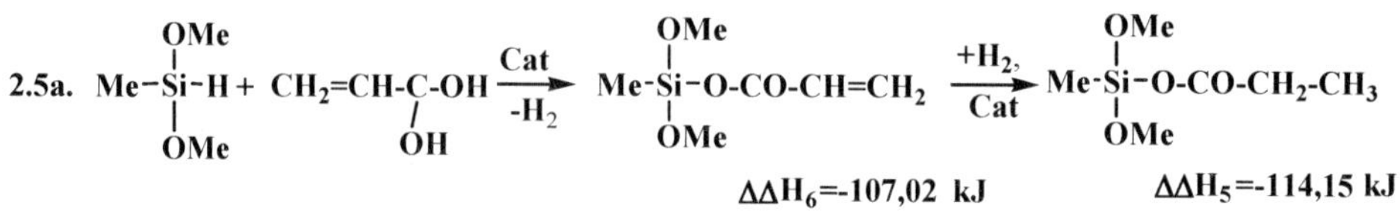

The structure of the product obtained by this way is identical to the structure of the product obtained by regrouping of intermediate product (scheme 2.4) according to the scheme

2.5a and they have the same values of calorific effects ($\Delta\Delta H=\Delta\Delta H_4+\Delta\Delta H_{4a}=\Delta\Delta H_6+\Delta\Delta H_5=-221{,}17$ kJ). The course of the reaction according to the schemes 2.4 and 2.4a compared with another course energetically is pro-fitable, the dehydrocondensation-dehydrogenation course of the reaction according to the scheme 2.5a is improbable, because during dehydrocondensation reaction released hydrogen is isolated from reaction mixture, that is fixed from reaction mixture during the process, but strong denial of course of the reaction by the scheme 2.5a is impossible. Besides, it is possible that sensitivity of the instrument and conditions of received spectra was insufficient for identification of all possible pro-ducts obtained with a low concentration.

On the scheme 3 one can see change of enthalpy of all reactions proceeding with various concurrent ways during the model hydrosilylation reaction of methyldimethoxysilane to methacrylic acid:

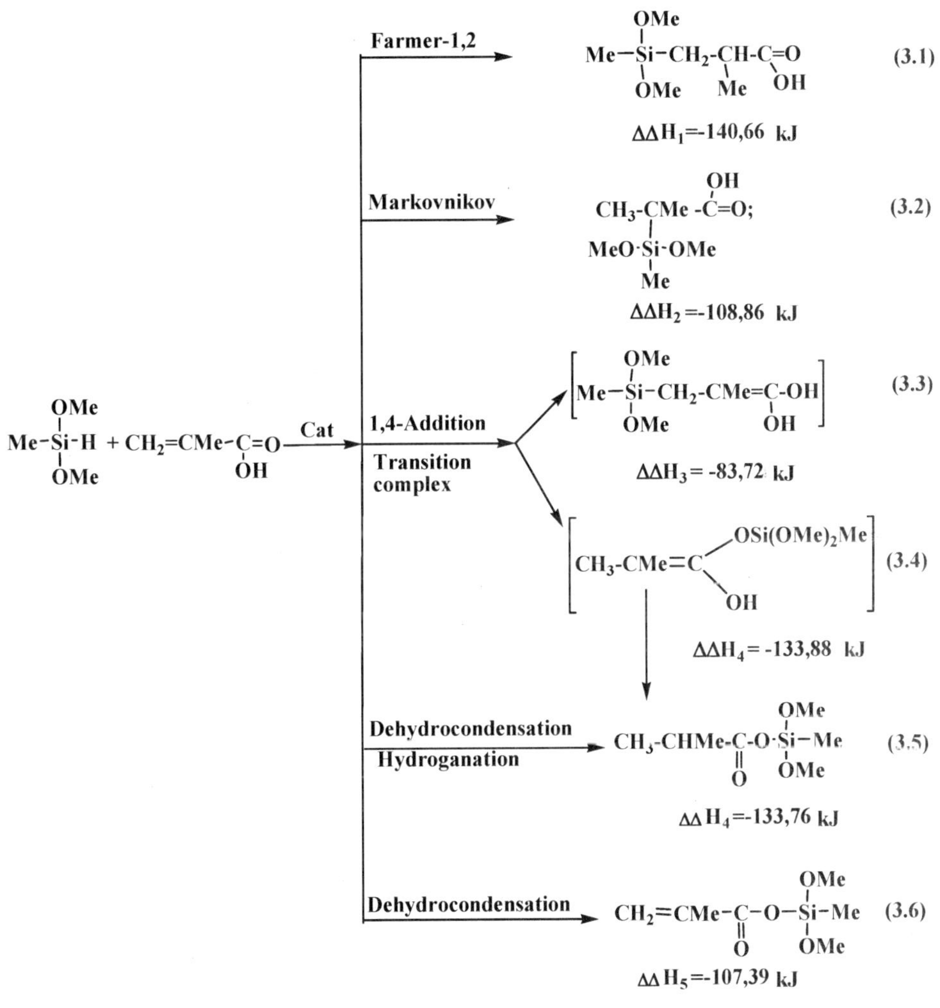

Scheme 3. Continued.

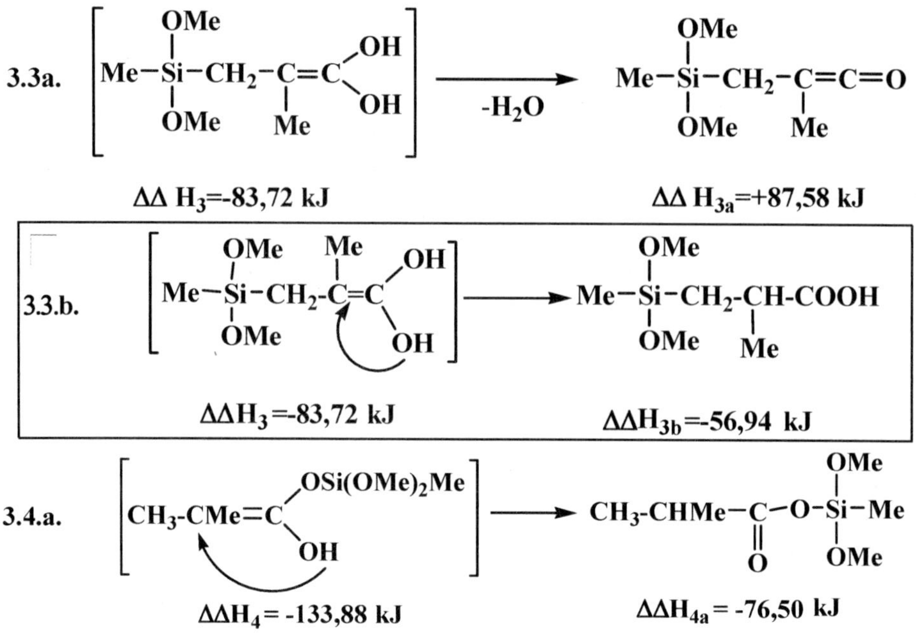

Scheme 3. Change of enthalpy of model reactions of hydrosilylation of methyldimethoxysilane to methacrylic acid.

On the assumption of heat of model reaction of methyldimethoxysilane with methacrylic acid (scheme 3), course of hydrosilylation according to Farmer rule (1,2-addition) thermodynamically is more profitable. The structure of the product, which is obtained by, regrouping of transition complex (1,4-addition, scheme 3.3) according to the Eltekov rule, scheme 3.3a, is similar to struc-ture of the product obtained by Farmer rule. Their heat of reaction $\{\Delta\Delta H_1=\Delta\Delta H_3+ +\Delta\Delta H_{3b}=$ [-83,72 kJ+(-56,94 kJ)]=-140,66 kJ} are same too. Formation of ketene group from transition complex ac-cording to the scheme 3.3a probable most of all, because its heat of reaction is equal to $\Delta\Delta H_{3a}=$ +87,58 kJ. In spite of this, as distinct from acrylic acid, in methacrylic acid obtaining of ketene group is proved by ^{1}H and ^{13}C NMR spectra data.

With respect to dehydrocondensation-dehydrogenation reaction of methyldimetho-xysilane with me-thacrylic acid which more comprehensive aspect represented as:

3.5a. 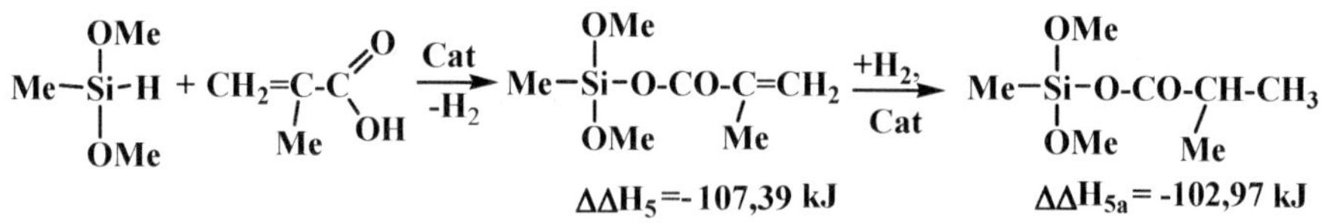

analogically the course of the reaction according to above-mentioned scheme 3.5 is improbable, though its exclusion is not possible.

By wide-angle X-ray analyses of oligomer IV[3] it was shown, that oligomer is one phase amorphous system with value of interchain distances $d_1 \approx 9{,}07$ Å.

So, at the first time it was studied hydride addition reactions of methylhydridesiloxane to acrylic and methacrylic acid in the presence of catalysts Platinum hydrochloric acid and Pt/C and differed from literature data it was shown, that besides hydrosilylation reactions (1,2- and 1,4-addition), de-hydrocondensation reactions takes place with formation various-linked thermoreactive oligomers.

REFERENCES

[1] Petrov A.D., Sadikh-Zade S.I., Filatova E.I. *Zhurn. Obshch., Khimii,* 1959, v.29, p.2936.

[2] Sadikh-Zade S.I., Petrov A.D. *Zhurn., Obshch., Khimii,* Chemistry and practical application siliconorganic compounds, part 1, Leningrad, 1958, p.212.

[3] Andrianov K.A., Abkhazava I.I., Khananashvili L.M., Bagratishvili G.D. *Zhurn., Obshch. Khimii,* 1973, v.43, №7, p. 1505.

[4] Andrianov K.A., Abkhazava I.I., Khananashvili L.M., Bagratishvili G.D., Doksopulo T.P., Chagelishvili V.A. *Zhurn., Obshch. Khimii,* 1973, v.41, №1, p. 606.

[5] Gverdciteli I.M., Doksopulo T.P., Chikovani E.I. *Zhurn., Obshch. Khimii,* 1975, v.47, №7, p. 352.

[6] Nitzsche S. *Angew. Chem.,* 1960, v. 72, No3, p. 112.

[7] Dolgov B.N. Chemistry and Practical Application of Organosilicon Compounds Central Library of Technical Information, Leningrad, 1958, v.1, p.18

[8] Nogaideli A.I., Tkeshelashvili R.Sh., Nakaidze L.I., Mukbaniani O.V. Works of Tbilisi State University, 1976, v.167, p. 69.

[9] Mukbaniani O.V., Khananashvili L.M., Esartia I.G., Khaduri S.D. *Intern. Journ. of Polymeric Materials,* 1994, v.24, №1, p. 131.

[10] Dolgov B.N., Kharitonov I.P., Vorokvov M.G. *Zhurn., Obshch., Khimii,* 1954, v.24, p. 861.

[11] Baines J., and Eaborn C. *J. Chem. Soc.,* 1955, p. 4023.

[12] Mukbaniani O.V., Scherf U., Gurgenidze G.N., Karchkhadze M.G., Meladze S.M., Kha-nanashvili L.M. *International Journal of Polymeric Materials,* 2001, v.48, No 3, p. 267.

[13] Mukbaniani O.V., Gurgenidze G.N., Meladze S.M., and Khananashvili L.M. Russian *Polymer News,* 2001, v.6, No4, p. 18.

[14] Dewar M.I.S., Zcobish E.G., Healy E.F., Stewart J.S. *J. Am. Chem. Soc.,* 1985, v.107, p. 3902.

In: New Topics in Monomer and Polymer Research
Editors: A. D'Amore and G. Zaikov, pp. 153-161 ISBN: 1-60021-436-3

Chapter 10

INFLUENCE OF 5,10,15,20-TETRAKIS-(3,5-DI-TERT-BUTYLPHENYL)PORPHYRIN AND ITS METAL COMPLEXES ON THE RADICAL POLYMERIZATION OF METHYLMETACRYLATE

Yu. B. Monakov, O. I. Koifman, R. H. Nasretdinovè, R. M. Islamova, Yu. I. Puzin and I.A. Ionova
Institute of Organic Chemistry, Ufa Scientific Center of RAS
pr. October 7, Ufa, 1450054 Russia
Ivanovo State University of Chemistry and Technology
Engels str., 7, Ivanovo, 153460, Russia
Bashkir State University
Frunze str., 32, Ufa, 450074, Russia
Ufa State Petroleum Technological University
Kosmonavtov str., 1, Ufa, 450062 Russia

ABSTRACT

The influence of 5,10,15,20-tetrakis-(3,5-di-tert-butylphenyl)porphyrin and its metal complexes on the radical polymerization of methylmetacrylate initiated by benzoyl peroxide has been studied. It was revealed that metal porphyrins form an initiating system with a peroxide initiator and increase the rate of the initial stage of polymerization. It was shown that addition of porphyrins leads to the formation of polymethylmetacrylate with lower molecular weight, enhanced thermostability and higher stereoregularity due to the increased presence of syndiotactic sequencies.

Complexes of porphyrins with transition metals are of considerable interest for macromolecular chemistry. Natural porphyrins play a central role in many fundamental biochemical systems containing biopolymers, e.g. in photosynthesis, in oxygen transfer, in the catalysis of metabolytic processes. In the last decade growing interest has been given to the synthesis and study of porphyrin-polymer systems with diverse supramolecular architecture and their prospective applications as nonlinear optical materials, optical sensors, gas-separation membranes, electron and photoelectron conductors, new catalytic materials[1]. Synthetic metalloporphyrins can improve thermostability of polymers[2] and exhibit a catalytic activity in macromolecular synthesis[3,4]. For example, it was shown that in the presence of cobalt[5,6] and aluminium[7,8] complexes of porphyrins the radical polymerization of acryl polymers could occur according to the "living polymerization" mechanism. Such polymerization mechanism leads practically to a monodisperse polymer which is characterized by a narrow molecular weight distribution (MWD) $M_W/M_N = 1.1$.

It should be noted that low solubility of porphyrin complexes in monomers has limited for a long time their study in radical polymerization. Nowadays this problem is solved since porphyrins with functional groups enhancing solubility became available. The study of behaviour of such compounds in polymerization of vinyl monomers is of considerable interest. In the present work the influence of additions of 5,10,15,20-tetrakis-(3,5-di-*tert*-butylphenyl)porphyrin (I) and its metal complexes (II-V) on the radical polymerization of methylmethacrylate (MMA), initiated by benzoyl peroxide (BP) at 60^0 C was studied. It was of particular interest to find out if investigated porphyrins form initiating systems with benzoyl peroxide and to establish their influence on the molecular-weight characteristics, thermostability and microstructure of the polymers obtained.

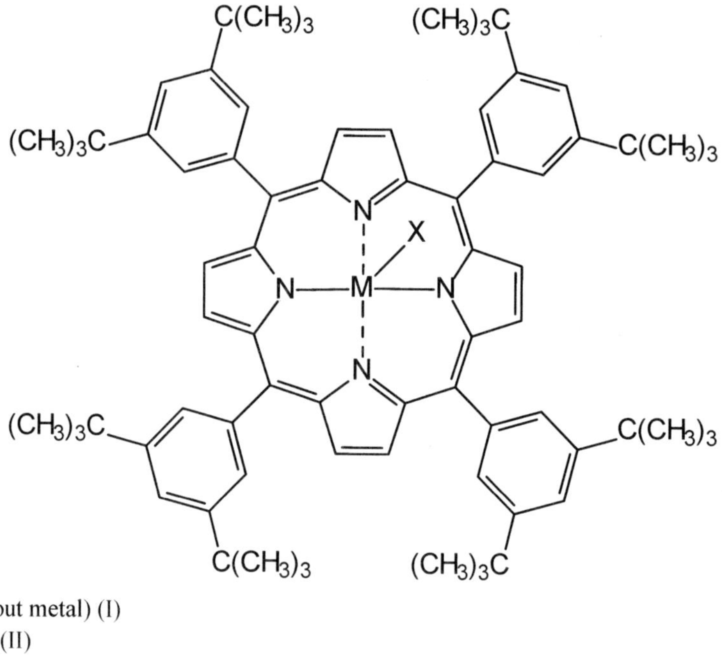

M = 2H; (without metal) (I)
M = Ti; X = O (II)
M = Zr; X = 2(AcAc) (III)
M = Fe; X = Cl^- (IV)
M = Co; X = CH_3COO^- (V)

EXPERIMENTAL SECTION

Methylmethacrylate (MMA) was purified from the addition of a stabilizer by shaking with 5-10% solution of KOH, followed by washing with water till neutrality, drying over $CaCl_2$ and double distillation under vacuum. Benzoyl peroxide was recrystallized three times from methanol (melting point 108°C with decomposition). 5,10,15,20-Tetrakis-(3,5-di-*tert*-butylphenyl)porphyrin (I) was obtained by the condensation of pyrrole and 3,5-di-*tert*-butylphenylbenzaldehyde[9]. Metal complexes (II –V) have been obtained from free base porphyrin and corresponding metal salts in dimethylformamide according to the literature procedure[10]. Kinetical parameters of polymerization have been obtained by dilatometric method [11].

Molecular-weight characteristics (weight-average and number-average molecular weights M_W and M_N and molecular-weight distribution (MWD) measured as M_W/M_N ratio) of polymethylmethacrylate (PMMA) have been determined using gel permeation chromatography on "Waters GPC 2000 System" liquid chromatograph using tetrahydrofurane as an eluent with flow rate 0,5 ml/min at 30^0 C. The column system was calibrated using polystyrole standards with narrow MWD ($M_w/M_n \leq 1,2$), using the universal Benoit dependence[12] and the equation connecting molecular weight of PMMA with characteristic viscosity in tetrahydrofurane[13]. Destruction of polymers has been studied on "Q – 1000" derivatograph (MOM, Hungary) in air with heating rate of 5 °C/min for the probe of 100 mg. The content of syndio- and isotactical sequencies in macromolecules has been determined by ^{1}H NMR spectroscopy[14] using Bruker AM-300 spectrometer in $CDCl_3$.

RESULTS AND DISCUSSION

As can be seen from Fig. 1, the polymerization of methylmethacrylate (MMA) initiated by benzoyl peroxide (BP) occurs more rapidly in the presence of porphyrin derivatives. The effect of porphyrin depends on its nature and among the investigated species compounds I, II and III exhibit the largest accelerating effect on the initial rate of polymerization process (Fig. 1).

For porphyrins IV and V the highest acceleration is observed at small conversion degree (below 3-4%), and then the rate is decreased and polymerization process occurs with the rate somewhat lower than the initial one. Such dependence is characteristic for nitrogen containing initiating systems with peroxides[15] and can be considered as an evidence for the formation of initiating systems BP-porphyrins.

Although the behaviour of porphyrin I and its metal complexes II-V is similar in many respects (Fig. 2), the central metal in the porphyrin core has a considerable effect on the polymerization process.

For initiating systems containing porphyrins I and II the rate of polymerization is increased with the raise of the porphyrin concentration in the range $0\text{-}1 \cdot 10^{-3}$ M. For the system containing V, the rate is increased up to $0.25 \cdot 10^{-3}$ M and then remains almost constant, while for porphyrins III and IV, the polymerization rate is decreased above $0.25 \cdot 10^{-3}$ M.

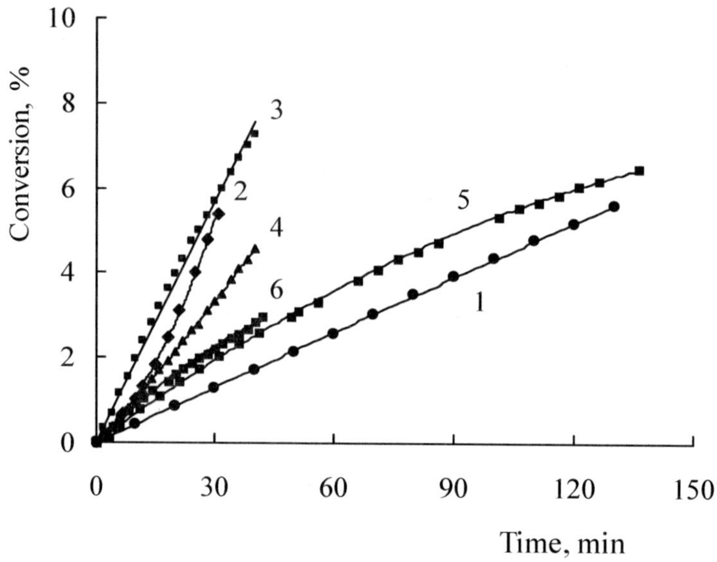

Figure 1. Kinetic plots of MMA polymerization of initiated with BP ($1 \cdot 10^{-3}$ M) without (1) and in the presence of porphyrins ($0.25 \cdot 10^{-3}$ M) - I (2), II (3), III (4), IV (5), V (6) at 60^0C.

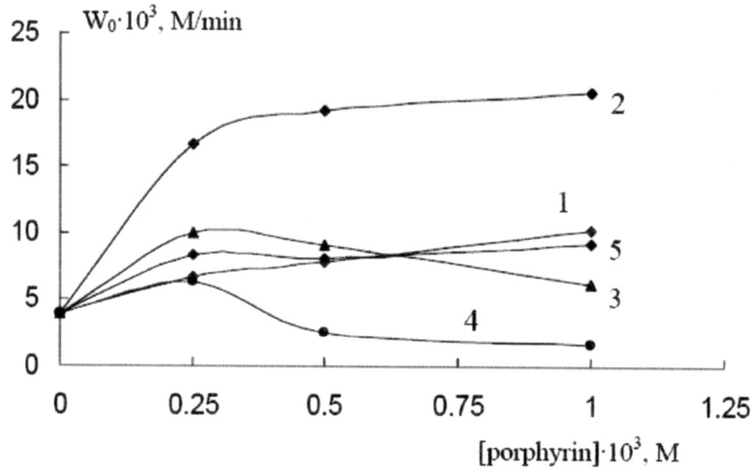

Figure 2. Dependence of the initial rate of MMA polymerization W_0 in the presence of BP ($1 \cdot 10^{-3}$ M) upon the concentration of porphyrin additions: I (1); II (2); III (3); IV (4); V (5). Temperature of synthesis is 60°C.

Fig. 3 displays the dependence of molecular weight (M_w) of polymethylmethacrylate (PMMA) obtained in the presence of initiating systems containing BP and porphyrins I – V upon the concentration of the addition.

Molecular weights of polymers obtained using initiating systems with BP and porphyrins I – V are decreased as compared to PMMA obtained in the absence of the porphyrin additions (Fig. 3). It is noteworthy, that there is a distinct relationship between the rate of polymerization and molecular weight of the polymer. Thus, for the systems containing BP and additions of porphyrins I or II the rate of polymerization is increased and molecular weights of PMMA are decreased. Such relationship is characteristic for the initiating systems containing the compounds exhibiting catalytic functions in respect to the initiator and was observed for the metallocenes – peroxides systems and some systems with nitrogen-bases[16,17]. One can suppose that in the present case porphyrins form the systems in which they catalyze the decomposition of benzoyl peroxide. For a system containing BP and porphyrin III the decrease of polymerization rate is accompanied by increase of M_w of PMMA. This is indicative that increase of the porphyrin concentration leads to formation of complexes with free radicals which are not active in polymerization and as a result the number of active centers of the chain propagation is decreased. For the systems with porphyrins IV and V the molecular weight of PMMA is decreased with the decrease of polymerization rate. This might be connected with the formation of non-active complexes with macroradicals of propagation, i.e. porphyrins IV and V being the chain transfer agents participate in the processes of chain termination. In the case of cobalt complex V it was not possible to obtain a polymer when the concentration of porphyrin was $1 \cdot 10^{-3}$ M.

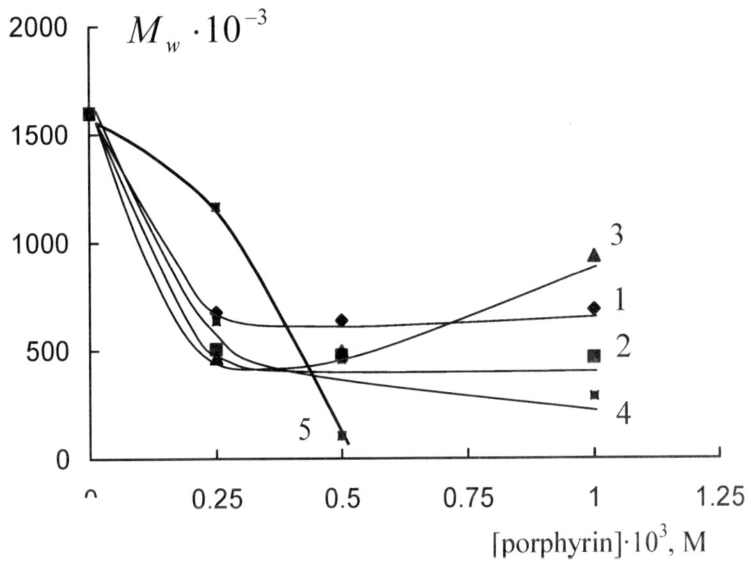

Figure 3. Dependence of M_w of PMMA upon the concentration of porphyrin I (1), II (2), III (3), IV (4), V (5). Conditions of polymerization - see Fig. 1.

The average number molecular weight M_n depends on the concentration of porphyrin in a similar way (Fig. 4).

The curves of MWD for the samples of PMMA obtained with initiator systems containing BP and porphyrins I – IV are shifted to a lower molecular weight region, but their pattern is practically the same as for PMMA obtained only in the presence of BP (Fig. 5). This corresponds to the polydispersity values which are not changed for the samples obtained with and without addition of porphyrin ($M_w/M_n = 1.8 \div 2.1$).

For the polymers obtained with the addition of Co-porphyrin complex V the polydispersity is increased as compared with initial PMMA from $M_w/M_n = 2$ to 3. This results in the corresponding changes of MWD curves pattern which not only shifts to the lower molecular weight region, but gain a bimodal character (Fig. 6).

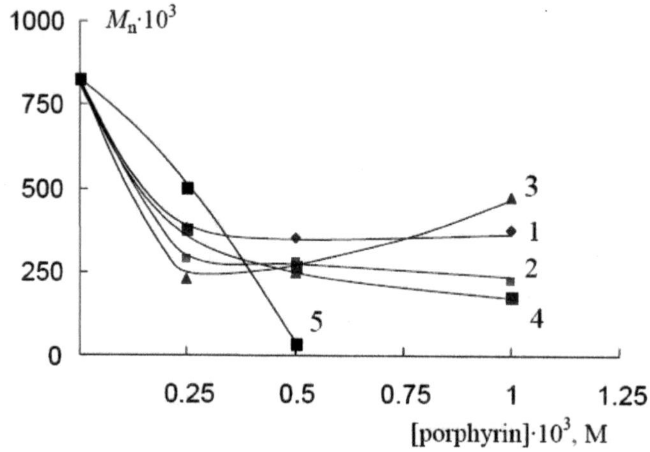

Figure 4. Dependence of M_n of PMMA upon the concentration of porphyrin. Initiating systems containing BP and I (1), II (2), III (3), IV (4), V (5). Conditions of polymerization - see Fig. 1.

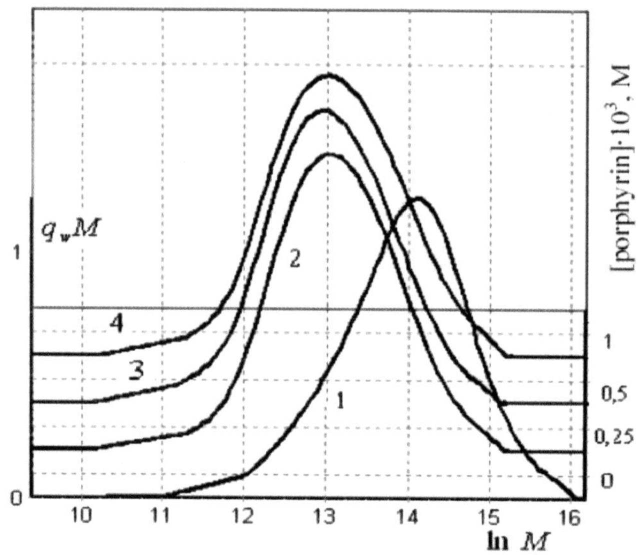

Figure 5. The MWD curves for PMMA obtained in the presence of initiating system BP – porphyrin I. Concentration of porphyrin I : 0 (1), $0.25 \cdot 10^{-3}$ (2), $0.5 \cdot 10^{-3}$ (3), $1 \cdot 10^{-3}$ (4).

The addition of porphyrins into a polymerizing system containing BP increases the thermostability of PMMA obtained (Table 1).

It is known that the decomposition of PMMA obtained by radical polimerization begins with the terminal unsaturated groups[18]. Antibathic character observed for the changes of the molecular weight and decomposition temperature allows to suppose that additions of porphyrins lead to the decrease of the number of unsaturated terminal groups in a forming polymer. This is also confirmed by DTG curves (Fig. 7).

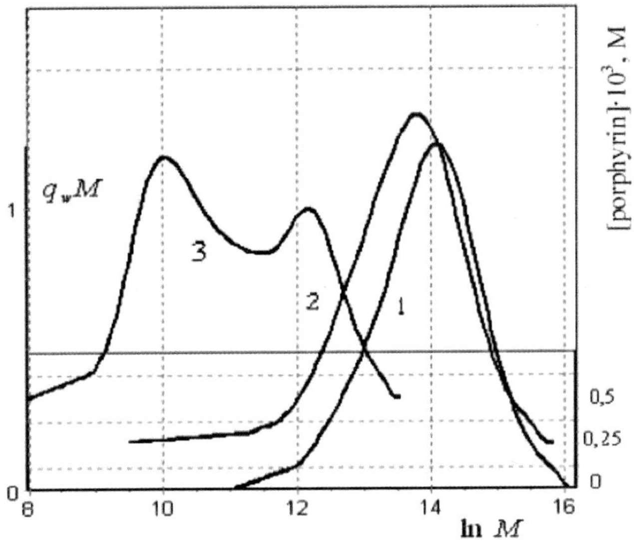

Figure 6. MWD curves for PMMA obtained in the presence of initiating system BP – porphyrin V. Concentration of porphyrin V: 0 (1), $0.25 \cdot 10^{-3}$ (2), $0.5 \cdot 10^{-3}$ M (3).

Table 1. Temperature of decomposition for PMMA obtained in the presence of BP and porphyrins I-V. $T_{pm} = 60^0C$, $[BP]=1 \cdot 10^{-3}$ M, $[porphyrin] = 0.25 \cdot 10^{-3}$ M.

Porphyrin	T_{decomp}. ^0C
-	224
I	266
II	273
III	264
V	255

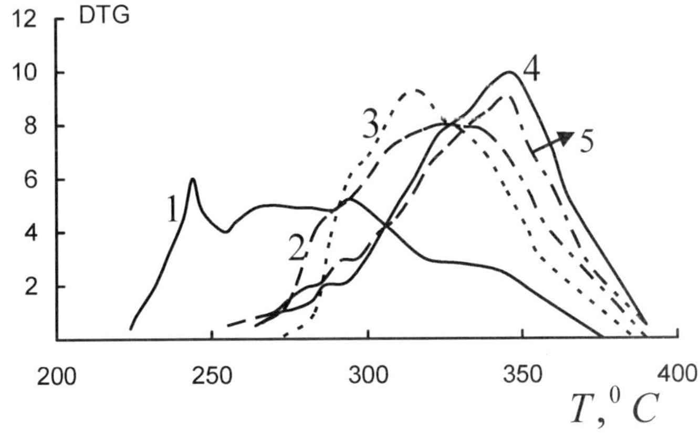

Figure 7. DTG curves of PMMA obtained in the presence of BP (1) and with addition pf porphyrins: I (2), II (3), III (4), V (5). Polymerization temperature Tpm = 600C, [BP]=1.10-3 M, [porphyrin] = 0.25.10-3 M.

As can be seen from Fig. 7 the DTG curve of PMMA obtained without the addition of porphyrins contains two peaks at 240 and 290 ^{0}C, in the presence of porphyrins the low temperature peak disappears. Therefore the addition of porphyrins has the effect on the lower temperature stage of the PMMA destruction, involving the splitting of the unsaturated terminal groups formed as a result of chain rupture due to disproportionation of macroradicals [18].

The addition of porphyrins in a polymerizing system has an influence also on the microstructure of a polymer (Table 2). In the presence of porphyrins the content of syndiotactic sequences is increased (Table 2). Such effect is also typical for metallocene[16,19] and nitrogen-base containing [17] initiating systems.

Table 2. Microstructure of PMMA obtained at 60°C with BP as an initiator in the presence of porphyrin additions. [BP] = $1{\cdot}10^{-3}$ M

Porphyrin	[Porphyrin]$\cdot 10^3$, M	Content of triades, %		
		syndio-	hetero-	iso-
-		56.0	42.0	2.0
IV	0.3	64.3	32.6	3.1
V	0.3	64.2	32.3	3.5

On the other hand, in the presence of porphyrin additions the content of isotactic sequences is also somewhat increased. This might be due to the formation of an intermediate from porphyrin and a propagation radical. The addition of the monomer unit to such intermediate occurs in a regular manner.

The results of the present study allow to conclude that the addition of porphyrins in a polymerizing system containing BP and MMA increases the rate of the initial stage of polymerization process resulting in a considerable decrease of the molecular weights of the PMMA. The polymers obtained are characterized by improved thermostability and higher stereoregularity.

This is evidently connected with the participation of porphyrin complexes with transition metals in the formation of new more effective initiating systems with a peroxide initiator. The presence of the metal increases strongly the influence of the porphyrin on the polymerization process and on the properties of the obtained polymer. On the other hand the behaviour of the complexes in polymerization is similar with the behaviour of the free base porphyrin. This indicates the substantial contribution of the porphyrin macrocycle itself in the formation of initiating systems.

ACKNOWLEDGEMENTS

This work was supported by Russian Foundation of Basic Research (grant Nr. 05-03-32087) and by Foundation supporting scientific schools (Grant NS-728.2003.3). The authors are thankful to Dr. V. A. Kraikin for his help in derivatographic analysis of polymers and Prof. A. S. Semeikin for synthesis of the metalloporphyrins.

REFERENCES

[1] Koifman O. I., Ageeva T. A. *Polymer Science, Ser. C.* 2004, 46(1), 49.

[2] (a) Smirnova G. N., Koifman O. I., Kirillova Ya. V., Snegireva F. P. Izv. Vyssh. Uchebn. Zaved., Khim. Khim. Tekhnol. 1990, 33(12), 125-6. (b) Markova L. V., Smirnova G. N., Korzhenevskii A. B., Koifman O. I. Izv. Vyssh. Uchebn. Zaved., Khim. Khim. Tekhnol. 1992, 35(11-12), 98-102. (c) Belokurova A. P., Aseeva R. M., Koifman O. I. *Izv. Vyssh. Uchebn. Zaved., Khim. Khim. Tekhnol.* 1993, 36(6), 97-100.

[3] Smirnov B.R. Catalysis of the chain transfer on monomer in radical polimerization. In: Porfiriny: spectroskopija, electrokhimija, primenenie (Porphyrins: spectroscopy, electrochemistry, application). Ed. Enikolopian N.S. Moscow: *Nauka,* 1987. P. 290-343.

[4] Aida T., Inoue S. Metalloporphyrins as catalysts for precision macromolecular systhesis. In: The Porphyrin Handbook. Vol. 6. Edrs Kadish K.M., Smith K.M., Guilard R. New York: Academic Press, 2000. P. 133-156.

[5] Korolyov G.V., Marchenko A.P. *Usp. Khim.* 2000, 69(5), 447.

[6] Wayland B.B., Basickes L., Mukerjee S. *Macromolecules.* 1997, 30, 8109.

[7] Aida T., Inoue S. *Macromolecules.* 1981, 14, 1162-1166.

[8] Aida T., Inoue S. *Acc. Chem. Res.* 1996, 29, 39-48.

[9] Semeikin A. S., Koifman O. I., Berezin B. D. *Khim. Geterotsikl. Soedin.* 1986, (6), 798-801.

[10] Smith K.M. Porphyrins and Metalloporphyrins. Amsterdam-London-New York: Elsevier. 1975.

[11] Gladyshev G.P. Polimerizatsija vinilovykh monomerov [Polymerization of Vinyl Monomers]. Alma-Ata: Nauka, 1964.

[12] Benoit H., Crubisic L., Rempp P. A. *J. Polym. Sci.* 1967, B5, 753.

[13] Styskin E.L. Prakticheskaja vysokoeffektivnaja zhidkostnaja khromatografia [Practical high-efficiency liquid chromatography]. Moscow: *Khimija,* 1986. 288 p.

[14] Shapiro Yu.E. *Bull. Magn. Reson.* 1985, 7(1), 27.

[15] Efremova E.P., Chikhaeva I.P., Stavrova S.D., Motov S.A., Pravednikov A.N. Doklady AN SSSR. 1981, 258(2), 354.

[16] Puzin Y.I., Yumagulova R.Kh., Kraikin V.A. *Europ. Polym. J.* 2001, 37(9), 1801.

[17] Puzin Y.I., Galinurova E. I., Monakov Y.B. *Vysokomol. Soedin. Ser. A.* 2002, 44(10), 318.

[18] Grassi N., Scott G. Destruction and stabilization of polymers. Moscow: *Mir,* 1988.

[19] Kolesov S.V., Yumagulova R.Kh., Prokudina E.M. *Vysokomol. Soedin. Ser. A.* 2003, 46(2), 324.

In: New Topics in Monomer and Polymer Research
Editors: A. D'Amore and G. Zaikov, pp. 163-169
ISBN: 1-60021-436-3

Chapter 11

RADICAL POLYMERIZATION OF METHYL METHACRYLATE IN THE PRESENCE OF N, N-DIMETHYL-N - (METHYLFERROCENYL)AMINE

Yu.B. Monakov*[*1], Yu.I. Puzin[1,2], A.V. Zaikina[1], I.I. Yarmukhamedova[1] and A.A. Fat'ykhov[1]

[1] Institute of Organic Chemistry of the Ufa' Reseach Center of RAS, Prospect Octyabrya 71, Ufa, 450054 Russia

[2] Ufa' State Petroleum Technological University, Kosmonavtove str. 1, Ufa, 450062 Russia

ABSTRACT

The free-radical polymerization of methyl methacrylate has been investigated in the presence of N, N-dimethyl-N - (methylferrocenyl)amine (FCAM). The mutual influence of ferrocenyl- and amine - groups on kinetic parameters of polymerization and on the microstructure of received polymers has been found. It is shown that poly-methyl methacrylate obtained in the presence of FCAM has mainly sindiotactic structure.

During radical polymerization the polymer is enriched by sindiotactic sequences due to insignificant difference in activation energy between sindio- and iso-addition of monomer to growing macroradical (it takes about 4 kJ/mol [1]). The quantity of heterotactic sequences in macromolecule chain increases with temperature thus the number of sindiotactic triads decreases.

By using some initiating systems one can influence to a number of polymer properties during radical polymerization including its microstructure. The polymerization in the presence of ferrocene – benzoyl peroxide system not only proceeds much faster, but also poly-MMA and polystyrene obtained have smaller molecular mass (MM), narrower molecular-mass distribution (MMD) and increased sindiotactic sequences content [2]. On the other hand, the rate of peroxides decomposition is considerably increased in the presence of amines in comparison with its decomposition in the absence of amines [3, 4]. The influence of heteroaromatic diketocarboxylic acid amine-groups on the polymer microstructure is so high that it is shown at lower concentration than peroxide and metallocenes [2, 5, 6]concentration.

However, the peroxides - amines systems is characterized by a significant process acceleration in the beginning of polymerization and inhibition of process with increasing monomer conversion [7]. It constitutes some problem to polymer syntheses. It is possible that the combination of metallocene and amine groups in the initiating system composition will allow to keep their stereo regulation function and also to increase polymerization rate.

In the given work the influence of N, N-dimethyl-N - (methylferroctnyl)amine (FCAM) and N, N-dimethyl-N-benzylamine on the methyl methacrylate (MMA) radical polymerization was investigated. FCAM contains ferrocenyl- and tertiary amine-groups:

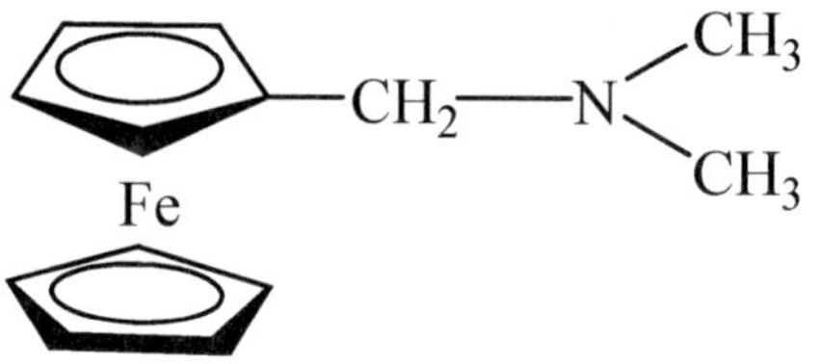

FCAM

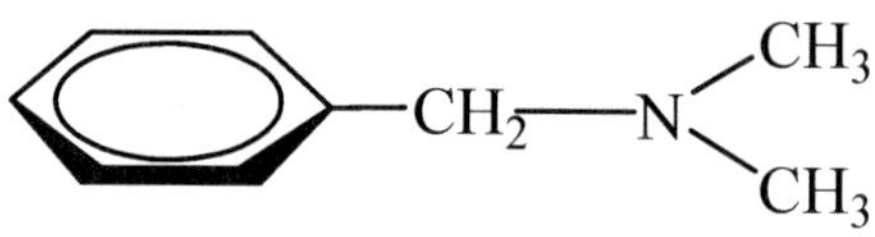

N, N-dimethyl-N-benzylamine

The initiating system with the participation of FCAM can show itself as system with ferrocene and as system with amines. Both functional groups can influence the radical polymerization simultaneously. That can have an effect both on the process and on the polymer microstructure.

EXPERIMENTAL PART

MMA were purified from the stabilizer by frequent washing with 5-10% KOH (in water) then with water. Monomer were dried over $CaCl_2$ and molecular sieves (3 Å) and distilled before use. Boiling points were 42 °C at 100 m. Benzoyl peroxide (melt temperature is 108 °C with decomposition) were recrystallized three times from methanol and dried to constant weight at room temperature in vacuum. FCAM was synthesized and purified by the technique [8]. N, N-dimethyl-N-benzylamine was received by Aldrich and was used without purification. The kinetic dependences of polymerization in bulk were studied dilatometrically [9].

The iso- and sindiotactic sequences content in macromolecules were determined by NMR ^{1}H on the device AM-300 "Bruker" [10]; polymer sample were obtained at low monomer conversion (~ 5 %). The solvent was $CDCl_3$.

Molecular weight of poly-MMA was determined by viscosimetry in benzene at 25±0.05 °C. Polymer samples for this purpose were obtained at low monomer conversion (~ 5 %), the

* Correspondence author: Yu.B. Monakov; e-mail: monakov@anrb.ru

polymer was precipitated in hexane, dried in vacuum at 40 °C up to constant weight. The equation [11] was used: [η]=KM$^{\alpha}$, где K=0.94×10^{-4}; α=0.76.

RESULTS AND DISCUSSION

The kinetic investigation of MMA polymerization initiated by the benzoyl peroxide - FCAM system has shown that this system differs from the benzoyl peroxide - ferrocene [2] and benzoyl peroxide - amines systems [3, 4, 7, 12] described earlier. The initial rate of polymerization was found to be dependent upon the concentration benzoyl peroxide and FCAM. The introduction of FCAM in the initiating system results in the increase of the initial rate of MMA polymerization in comparison with the polymerization rate without the additive (table 1). The orders with respect on the initiator were determined (table 1). The total order is close to 0.5 that testifies to radical type initiating system formation. The similar value was obtained earlier for ferrocene - benzoyl peroxide system [2]. The estimated activation energy of MMA polymerization initiated by the benzoyl peroxide - FCAM system is equal 45.7±2.5 кJ/mol and close to activation energy for benzoyl peroxide - ferrocene system (48.3±1.5 кJ/mol [2]). The kinetic investigation of MMA polymerization in the presence of benzoyl peroxide - N, N-dimethyl-N-benzylamine initiating system has shown that the order in respect on N, N-dimethyl-N-benzylamine is equal 0,1 (60 °C), 0 (45 °C), -0,1 (30 °C), and the order on benzoyl peroxide in this system is 0,4 (60 °C), 0,37 (45 °C), 0,27 (30 °C). The activation energy of polymerization with participation of N, N-dimethyl-N-benzylamine is 130 ± 2.5 kJ/mol. It exceeds the value for usual MMA radical polymerization initiated by benzoyl peroxide. The ferrocenyl function group possibly plays the determining role at the polymerization in the presence of benzoyl peroxide - FCAM system.

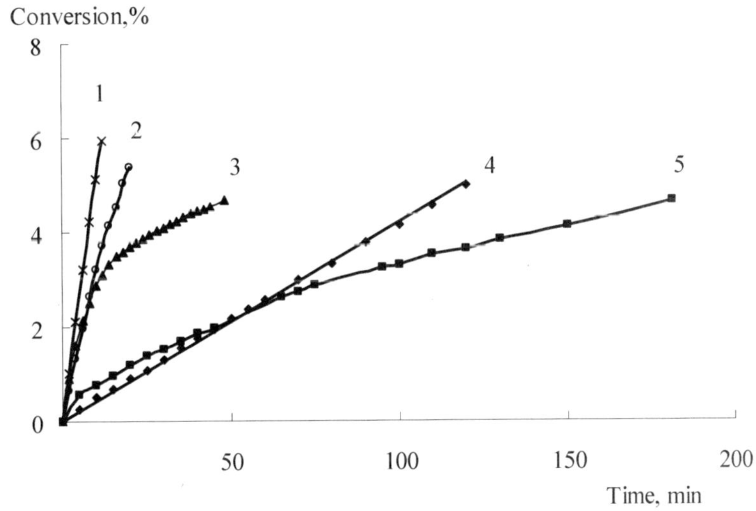

Figure 1. Dependent of MMA conversion vs polymerization time at 60 °C. [benzoyl peroxide]=1×10^{-3} mol/l, [ferrocene]=1×10^{-3} mol/l (1); [benzoyl peroxide]=1×10^{-3} mol/l, [FCAM]=1×10^{-3} mol/l (2); [benzoyl peroxide]=1×10^{-3} mol/l, [FCAM]=4×10^{-3} mol/l (3); [benzoyl peroxide]=1×10^{-3} mol/l (4); [benzoyl peroxide]=1×10^{-3} mol/l, [N, N-dimethyl-N-benzylamine]=4×10^{-3} mol/l (5).

Despite rate decreasing the MMA polymerization in the presence of FCAM is not finished in low degrees of monomer conversion and characterized by gel - effect. The increase of additive concentration results in increase of time for limiting monomer conversion achievement (fig. 2).

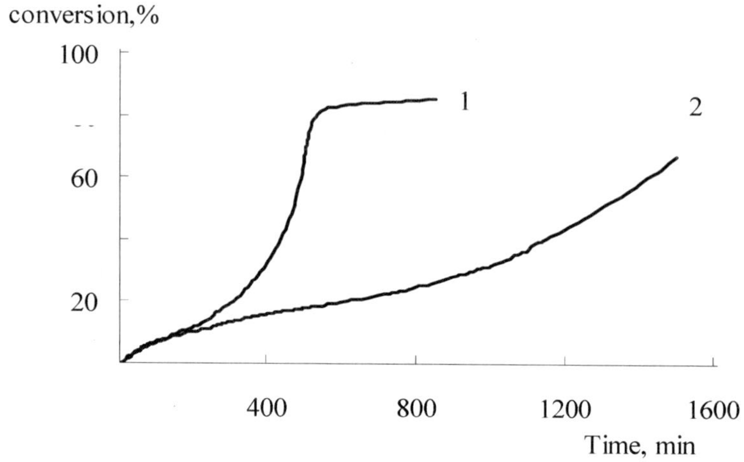

Figure 2. Kinetic plots of MMA polymerization in the presence of FCAM at 45 °C. [FCAM]×103=0.5 (1) и 1.0 (2) mol/l, [benzoyl peroxide]=1×10-3 mol/l.

So, both ferrocenyl- and amine- groups of FCAM affect polymerization of MMA and the amine- group inhibiting action increases with increase of additive concentration. The ferrocenyl- group playes determining role in initial monomer conversion. The kinetic parameters of polymerization are close to parameters which are characteristic for systems with ferrocene [2].

Poly-MMA sample synthesized in the FCAM presence has lower average degree of polymerization (P_n) (table 1). Such influence on polymers MM is typical of metallocenes [2].

Poly-MMA obtained in the presence of FCAM has mainly sindiotactic structure (table 2) though the isotactic sequences is also observed. It is found that heterotactic sequences amount increases with temperature increase and the sindiotriads content decreases. It is important to note that the microstructure of polymer chains practically does not depend on monomer conversion.

In fig. 3 the poly-MMA microstructure - synthesis conditions plots is shown. The data on the ferrocene influence to triads composition of polymer obtained in simulated condition [6] are given too. The presence of aminogroups seems to promote isotactic sequences formation and create certain obstacles to sindiotactic connection during chain propagation. On the contrary, the presence of ferrocene results in sindiotactic sequences formation.

In a fig. 1 the kinetic curves of MMA radical polymerization at 60 °C initiated by benzoyl peroxide in the presence of FCAM and N, N-dimethyl-N-benzylamine are shown. The introduction of FCAM to the initiating system results in increasing the initial rate of polymerization. However, then it is reduced and reached constant value. This feature is shown mostly when [FCAM] > [benzoyl peroxide]. In this case the polymerization rate decreasing occurs at rather lower monomer conversion (fig. 1, curves 3 and 4). Such character of polymerization is typical for peroxide - amines systems [12] and is observed for benzoyl peroxide - N, N-dimethyl-N-benzylamine system.

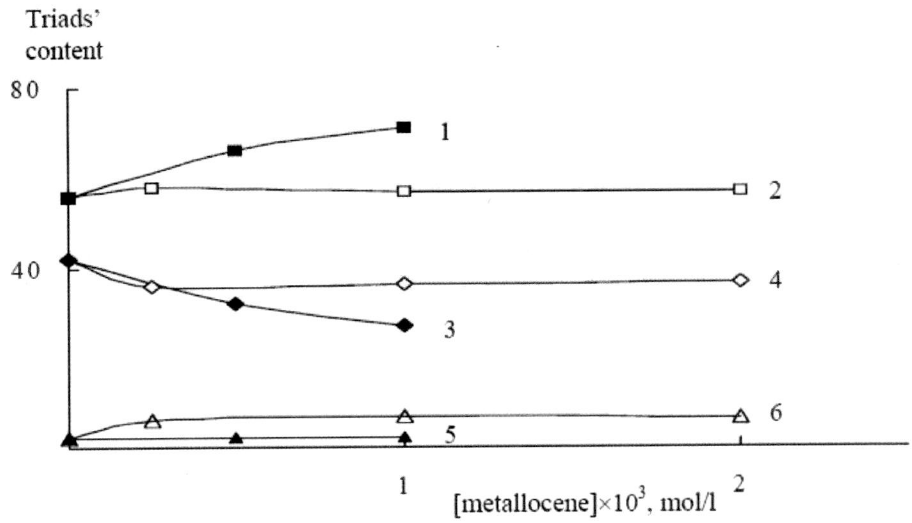

Figure 3. Plots of sindiotactic (1, 2), heterotactic (3, 4) и isotactic (5, 6) triads content of poly-MMA obtained at 60 °C (б) in the presence of 1×10^{-3} mol/l benzoyl peroxide vs concentration ferrocene (1, 3, 5) or FCAM (2, 4, 6).

Table 1. Polymerization of MMA in the presence of FCAM at 60 °C (initiator - benzoyl peroxide 1×10-3 mol/l)

T, °C	[FCAM]$\times10^3$, mol/l	$W_0\times10^3$, mol/l× min	$P_n \times 10^{-3}$	Order on FCAM	Order on benzoyl peroxide
60	-	3.8	17.1	0.22	0.21
	0.25	24.1	3.4		
	0.5	27.8	-		
	1.0	32.7	2.5		
	2.0	38.1	3.5		
	4.0	37.2	4.0		
45	-	1.1	22.8	0.30	0.33
	0.25	11.3	3.6		
	0.5	9.7	--2,4		
	1.0	17.0	2.4		
	2.0	18.5	2.4		
30	-	0.3	25.4	0.34	0.29
	0.25	3.4	2.9		
	0.5	5.9	-		
	1.0	6.0	3.2		
	2.0	7.4	3.7		
0	0.25	0.3	6.7	0.41	0.04
	0.5	0.6	6.3		
	1.0	1.3	4.5		
	2.0	0.7	5.4		

Unfortunately, the data on poly-MMA microstructure obtained in the presence of benzoyl peroxide - amines systems hasn't been found out in the literature. There are data about sindiotactic poly-MMA synthesis in the presence of benzoyl peroxide with cethylpiridinechloride systems [13] and with 3,6 - bis - (o- carboxybenzoyl) -N-isopropylcarbasol [5]. In both cases the isotriads content is insignificant. Macromolecules possibly is enriched by sindiotactic sequences in the presence of amine-groups without other function groups.

Ferrocene is known to accelerate benzoyl peroxide decomposition [2]. Metallocenes formes complexes with carbonyl groups of benzoyl peroxide [6]. It can be expected that such complex with FCAM is more stable than benzoyl peroxide complex with individual ferrocene and amine. It can affect stereo regulation of polymer chain.

So, simultaneous presence of ferrocenyl- and amine - groups decreases the sindiotriads content and increases isotactic sequences formation. It points to the mutual influence groups on activity of each other when they interractes with benzoyl peroxide.

Table 2. Date on microstructure of poly-MMA synthesized in the presence of FCAM

T, °C	[benzoyl peroxide] $\times 10^3$, mol/l	[FCAM] $\times 10^3$, mol/l	Triads content, %		
			sindio-	hetero-	iso-
60	1.0	---	56.1	42.0	1.9
	1.0	0.25	58.0	35.8	6.2
	1.0	1.00	56.9	36.4	6.7
	1.0	2.00	56.8	36.8	6.4
	0.5	1.00	60.0	33.8	6.2
	2.0	1.00	59.0	34.3	6.7
45	1.0	---	65.2	32.4	2.4
	1.0	0.50	63.7	33.4	2.9
	1.0	1.00*	62.0	34.4	3.6
	1.0	1.00**	62.2	34.5	3.3
	1.0	2.00	63.6	33.4	3.0
30	1.0	---	65.6	30.5	3.9
	1.0	1.00	68.3	29.7	2.0
20	1.0	1.00	69.4	28.8	1.8
0	1.0	0.50	72.8	26.0	1.2
	1.0	2.00	71.6	26.8	1.6

*conversion 40% и **80%

ACKNOWLEDGEMENTS

This work is supported by Russian Fund for basic researches (no. 05-03-32087a) and Fond of scientific school support (grant SS-728.2003.3). E-mail addresses: monakov@anrb.ru, puzin@anrb.ru

REFERENCES

[1] Encyclopedia of polymers. M.: *Soviet Encyclopedia,* 1974. V. 2. P. 205.
[2] Puzin Y.I, Yumagulova R.Kh., Kraikin V.A. // *Eur. Polymer J. V.* 37. 2001. P. 1801-1812.
[3] Chalt'ikyan O.A., Atanasyan E.N., Sarkisuyan A.A. // *Doklady Akademii Nauk Arm.S.S.R.* 1952. V. 15. P. 41.
[4] Milutinskaya R.I., Bagdasar'an Kh.S. // *Zhurnal phis. khimii.* 1960. V. 34. № 1. P. 405.
[5] Puzin Yu.I., Galinurova I.I., Kuznecov S.I., Fat'ikhov A.A., Monakov Yu.B. // *V'isokomolec. soed. A.* 2002. V. 44. № 10. P. 1752.
[6] Puzin Yu.I., Procudina E.M., Yumagulova R.Kh., Muslukhov R.R., Kolesov S.V. // Doklady Akademii Nauk. 2002. V. 386. № 1. P. 69.
[7] Pravednicov A.N., Stavrova S.D., Chikhacheva I.P., Ephremova E.P., Eremina M.G., Budanova E.N. // *Plasticheskie mass'i.* 1980. № 12. P. 10.
[8] Sintez'i organicheskich preparatov. Translation from English by Plate A.F. Moscow: *Mir,* 1964. V. 12. P. 192.
[9] Gladyshev G.P. Vinyl monomers polymerization. Alma-Ata: Izdatel'stvo Nauka, 1964.
[10] Shapiro Yu.I. // *Bull. Magn. Reson. 1985*. V. 7. №1. C. 27.
[11] Raficov S.R., Pavlova S.A., Tverdokhlebova I.I. Methods determination of polymer molecular weight and polydispersity. Moscow: Izdatel'stvo Akad Nauk S.S.S.R., 1962.
[12] Epimakhov Yu.K., Galibei V.I., Khudyakova U.A. // *V'isokomolec. soed. B.* 1985. V. 27. № 6. P. 464.
[13] Trubicina S.N., Margaritova M.F., Medvedev S.S. // Doklady Akademii Nauk S.S.S.R. 1966. V. 166. № 2. P. 381.

In: New Topics in Monomer and Polymer Research
Editors: A. D'Amore and G. Zaikov, pp. 171-176 ISBN: 1-60021-436-3

Chapter 12

THE EFFECT OF C_{60} FULLERENE ON RADICAL (CO)POLYMERIZATION OF ALLYMETACRYLATE AND ALLYLCHLORIDE

Yu.B. Monakov, R.Kh. Yumagulova, Yu.N. Biglova, R.F. Talipov, S.V. Kolesov
Institute of Organic Chemistry, Ufa Scientific Center of RAS
pr. October, 71; Ufa, 450054 Russia
Bashkir State University
Frunze str., Ufa, 32450074, Russia

Works on the synthesis and research into fullerene-containing polymers (FP) are paid particular significance in the chemistry of fullerenes. Radical polymerization is seen as especially attractive with such aims in view due to relative simplicity of the process to be realized, availability and variety of starting substances. The ways of obtaining fullerenes by vinyl monomer-based radical polymerization have been mainly discussed [1 - 5]. However up to date there are no reports on synthesis of these high-molecular compounds from the allyl order monomers. Interest is taken in the studies of the effect of fullerenes on the polymerization of the monomers containing simultaneously both vinyl and allyl double bonds, as well as the compounds having only allyl group. We have been first to do research into radical (co)polymerization of allyl ester of metacrylic acid (AMA) and allyl chloride (AC) in the presence of C_{60} fullerene.

The polymerization was carried out in mass at the temperature of 60°C. Fullerene C_{60} was used as a saturated solution in *o*-dichlorobenzene. The polymerization kinetics has been studied by dilatometry. The process was initiated with benzoyl peroxide at the constant concentration of 2.0×10^{-2} mol/L.

The kinetic dependences of the reactions of allyl and methyl esters of metacrylic acid in the absence of C_{60} fullerene can be correlated (Fig.1, curves 1 and 2). Although it can be noted that allylmetacrylate is polymerized at a faster rate. It is obviously due to the well-known fact of parallel cross-linking process along side allyl groups [6], contributing to the growth of contraction factor, that being perceived as the rise in the reaction rate. Insoluble cross-linked AMA polymers are produced at the conversion already ≥ 3-5% (within 15-20 minutes). The presence of fullerene considerably affects the observed kinetic functions. In radical polymerization fullerene is known to act as an inhibitor [7]. But at one and the same concentration of C_{60} the slowdown of allyl ester polymerization unlike that of methyl one is much more pronounced (Fig. 1, curves 4 and 5). In this case the degree of AMA conversion into soluble polymer goes up to 20%. The drop in the fullerene contents in the reaction mixture leads to a greater proportion of insoluble AMA polymer.

In copolymerization of allyl and methyl esters in the absence of fullerene the rise in the molal portion of AMA results in a higher reaction rate (Fig. 2, curves 1a and 2a). The increase in molal portion of allyl ester in the initial mixture at the fixed concentration of C_{60}, reversely, results in a lower process rate (Fig. 2, curves 1b and 2b). Copolymers obtained in the absence of fullerene are insoluble cross-linked products at the conversion of already 9% with the proportion of AMA in the reaction mixture of 10 mol %. The presence of C_{60} produces soluble samples until the conversion degree of ≤ 28%. This means that fullerene in the reaction mixture excludes cross-linking along side allyl groups, that leading to the growth of contraction factor.

The peculiarities of reactions of AMA homo- and copolymerization are due to the presence of two reaction centres: vinyl and allyl bonds in the molecule of this compound. The monomer, that contains only allyl group (allyl chloride having been employed as the former) undergoes no homopolymerization in the conditions studied. In copolymerization with methyl metacrylate (MMA) in the absence of fullerene the build-up of molal portion of low-active AC causes the drop in the process rate (Fig. 2, curves 3a and 4a). In the presence of C_{60} the rise in the proportion of allyl monomer in the starting mixture is accompanied by increase in the reaction rate (Fig. 2, curves 3b and 4b). The above fact is still another proof of the interaction of allyl radicals with C_{60} fullerene. Copolymers of AC with MMA are substances of good solubility at any monomer ratios.

The ultraviolet spectroscopy confirmed fullerene presence in polymer compositions. Free fullerene is characterized by two maximums of absorption at wavelengths λ_{max} = 260 nm and λ_{max} = 330 nm. In the spectra of AMA homopolymer solutions (chloroform is a solvent) obtained in the presence of fullerene an absorption strip appears at λ_{max} = 450 nm, which is caused by chemically bound fullerene (Fig. 3). The same absorption maximum is found in the copolymers. The proportion of fullerene in the synthesized under the same conditions polymer samples amounts to 0.2 mol/mol and 0.006 mol/mol for allyl and methyl esters respectively. Therefore, in polymerization of allyl monomer the quantity of fullerene joining polymer chains increases by an order of magnitude.

The available experimental data may apparently be interpreted as the result of the change in the reactivity of the fullerene molecule after the first addition reactions, i.e. as the display of regioselective effect, which characteristic of double bonds of C_{60}. The disruption of total π-electron system results in the formation of sites of greater activity on the fullerene [8], those sites are being likely to react with even low-active allyl radicals.

Thus, in the process of radical (co)polymerization of allyl ester of metacrylic acid and allyl chloride the interrelation of allyl radicals with C_{60}-fullerene has been first discovered. This fact excludes cross-linkage of macrochains, leading to a greater degree of the monomer conversion in soluble polymer and to a higher extent of the reaction products being functionalized by fullerene as compared to vinyl polymer. The discovered effect can enlarge the field of applying allyl monomers to synthesize new fullerene polymers.

The authors are thankful to Dr. S.I. Kuznetsov for his help in UV-spectra analysis of polymers.

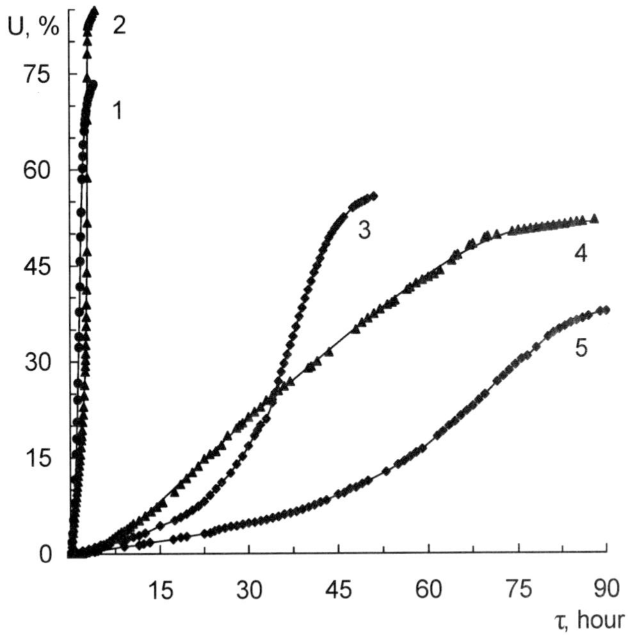

Figure 1. Conversion as a function of time in allymetacrylate (*1, 3, 5*) and methylmetacrylate (*2, 4*) in the presence of C_{60} fullerene. $[C_{60}]\times10^3$ mol/L: *1, 2* – 0; *3* – 2.25; *4, 5* – 4.5.

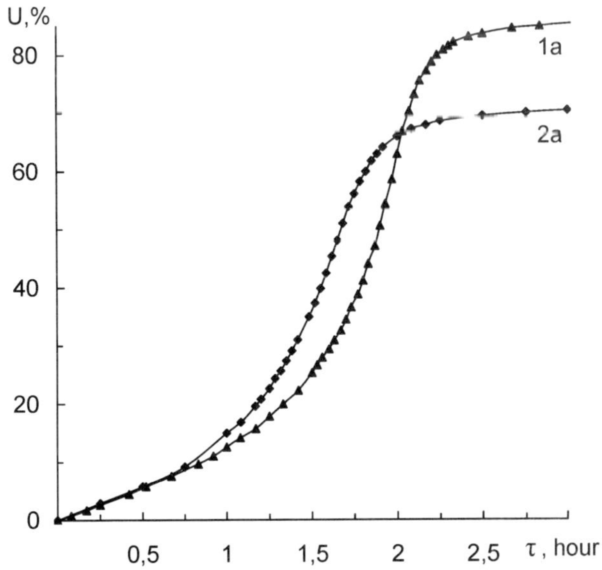

Figure 2. Continued

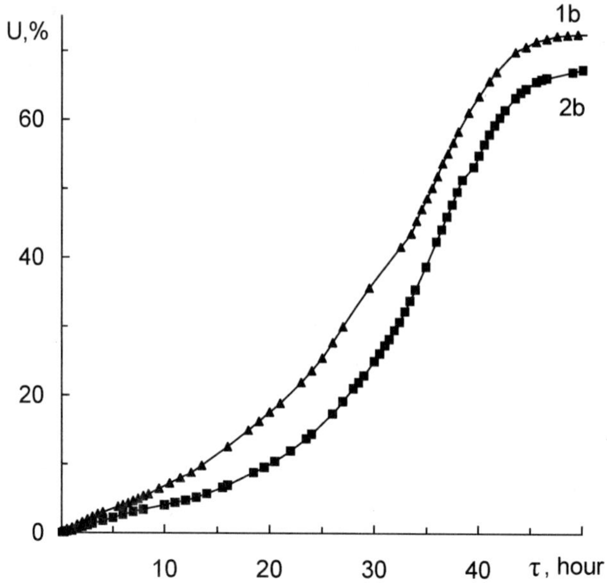

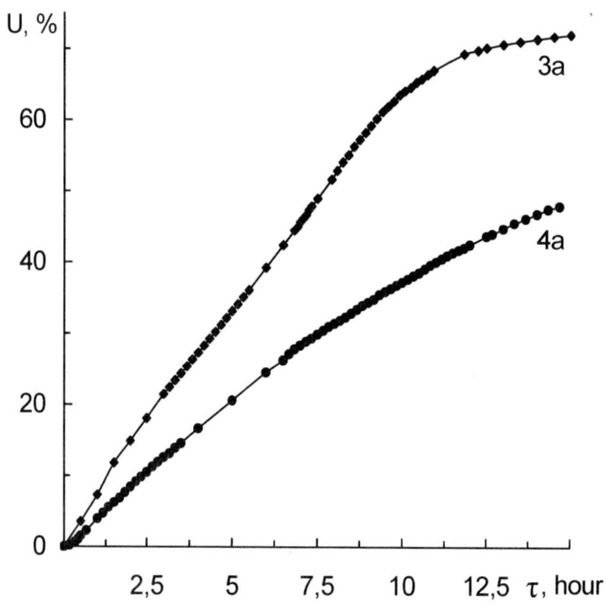

Figure 2. Conversion as a function of time in copolymerization of allymetacrylate (*1*, *2*) and allyl chloride (*3*, *4*) with methylmetacrylate in the presence of C_{60} fullerene. $[C_{60}]\times10^3$ mol/L: a) – 0; b) – 2.25. The monomer ratio (mol. portions): AMA (AC) : MMA = 0.3:0.7 – *1* (*3*); AMA (AC) : MMA = 0.5:0.5 – *2* (*4*). (Figure continues onto next page).

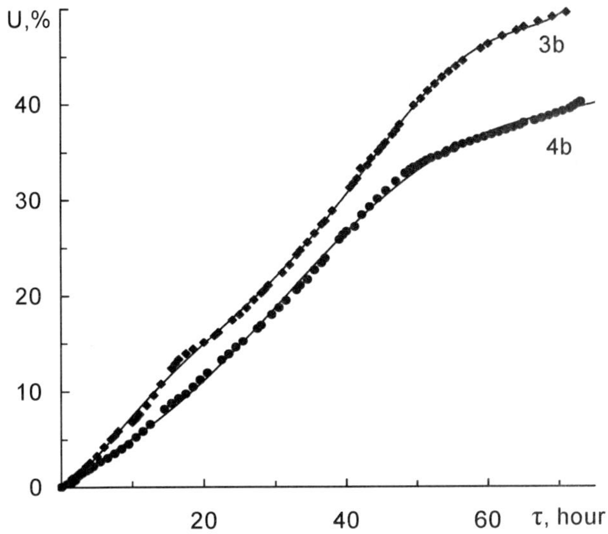

Figure 2. (Continuation) Conversion as a function of time in copolymerization of allymetacrylate (*1*, *2*) and allyl chloride (*3*, *4*) with methylmetacrylate in the presence of C_{60} fullerene. $[C_{60}]\times10^3$ mol/L: a) – 0; b) – 2.25. The monomer ratio (mol. portions): AMA (AC) : MMA = 0.3:0.7 – *1* (*3*); AMA (AC) : MMA = 0.5:0.5 – *2* (*4*).

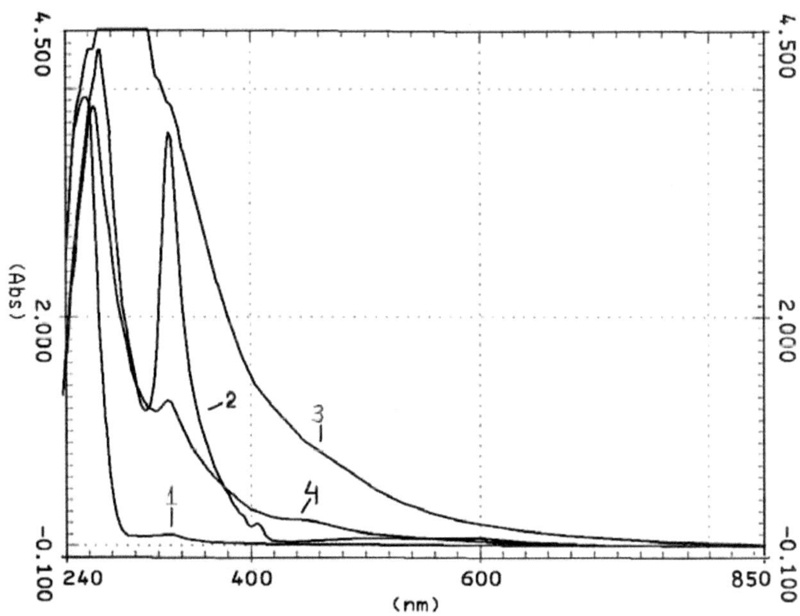

Figure 3. UV-spectra of the starting substances and the reaction products: *1* – AMA; *2* – C_{60}; *3* – PAMA; *4* – copolymer of AMA with MMA.

REFERENCES

[1] Camp A.G., Lary A., Ford W.T. // *Macromolecules.* 1995. Vol.28. №23. P.7959-7961.

[2] Ford W.T., Graham T.D., Mourey T.H. // *Macromolecules.* 1997. Vol.30. № 21. P.6422-6429.

[3] Sun Y.-P., Lawson G.E., Bunker C.E., Jonson R.A., Ma B., Farmer C., Riggs J.E., Kitaygorodskiy A. // *Macromolecules.* 1996. Vol.29. №26. P.8441-8448.

[4] Cao T.,Webber S.E. // *Macromolecules.* 1996. V.29. №11. P.3826-3830.

[5] Biglova Yu.N., Sigaeva N.N., Talipov R.F., Monakov Yu.B. // *Oxidation communications.* 2005. Vol.28. №4. P.753-798.

[6] Vorob'eva A.I. // Chemistry and physical chemistry of highmolecules compounds. Ufa: BFAN SSSR. 1987. P.32-42.

[7] Karpasheva G.P. // Vysokomolekulyarnye soedineniya. *Ser.C.* 2000. Vol.42. №11. P.1974-1999.

[8] Melenevskaya E.Yu., VinogradovaL.V., Litvinova L.S., Kever E.E., Shibaev L.A., Antonova T.A., Bykova E.N., Klenin S.I., Zgonnik V.N. // *Vysokomolekulyarnye soedineniya. Ser.A.* 1998. Vol.40. №2. P.247-254.

INDEX

A

B

I

J

K

L

M

N

O

P

Q

R

S

T

U

V

W

X

Y

Z